T0135820

Adaptive Polynomial Tabulation:
A Computationally Efficient Strategy
For
Complex Kinetics

This dissertation is approved by the Faculty of Environmental Sciences and Process Engineering at the Brandenburg University of Technology Cottbus in partial fulfillment of the requirement for the award of the academic degree of Doctor of Engineering in Process Engineering

by

Ngozi Ebenezer, M.Sc.

from Umuele, Njaba L.G. A, Imo State, Nigerian, born in Buea, Cameroon

Reviewer: Prof. Dr-Ing. Fabian Mauss (Thesis Supervisor)
Reviewer: Prof. Dr.-Ing. Dominique Thévenin

Date of oral examination: 18th December 2013

Bibliographic information published by the Deutsche Nationalbibliothek

The Deutsche Nationalbibliothek lists this publication in the Deutsche
Nationalbibliografie; detailed bibliographic data are available
in the Internet at http://dnb.d-nb.de .

ISBN 978-3-8325-3739-5

Logos Verlag Berlin GmbH
Comeniushof, Gubener Str. 47,
10243 Berlin
Tel.: +49 (0)30 42 85 10 90
Fax: +49 (0)30 42 85 10 92
INTERNET: http://www.logos-verlag.de

Adaptive Polynomial Tabulation:
A Computationally Efficient Strategy
For
Complex Kinetics

Von der Fakultät für Umweltwissenschaften und Verfahrenstechnik der Brandenburgischen Technischen Universität Cottbus zur Erlangung des akademischen Grades eines Doktor-Ingenieurs genehmigte Dissertation

vorgelegt von

Ngozi Ebenezer, M. Sc.

Aus Umuele, Njaba L.G.A, Imo State, Nigeria, geboren Buea, Kamerun

Gutachter: Prof. Dr-Ing. Fabian Mauss
Gutachter: Prof. Dr.-Ing. Dominique Thévenin

Tag der mündlichen Prüfung: 18 Dezember 2013

iii

Acknowledgements

I wish to thank my God, the Almighty God and my Savior Jesus Christ for making this 'engracement' possible. Mark 9:23, 'All things are possible to him that believeth'.

I am grateful to the supervisor of this thesis, Prof. Fabian Mauss, first for accepting to supervise me and guiding me through this adventure. His knowledge of combustion makes him one of the finest geniuses in our time. I appreciate the assistance of Prof. Ingemar Magnusson and Dr. Edward Blurock for serving as assistant supervisors for this work. I wish to thank Dr. Blurock for introducing me to programming, pointers and Principal Component Analysis. I wish to give special thanks to Prof. Thévenin for his patience in reviewing this thesis as well as his suggestions. I am grateful to the following members of the examination panel: Prof. Goerg Bader, Prof. Ebehard Schaller and Dr. Klaus Keuler. I heartily appreciate Prof. Klaus Schnitzlein and Dr. Dagmar Stephan for doing all the administrative work behind the scenes for the defense and publication of this thesis. I salute Mrs Printschitsch for her patience in receiving and responding my calls and emails.

I wish to thank Mrs Uhlemann, Mr Brandt, Mr Roth, Ms Kasper, Dr Tabet, Vivien, Amruta, Sayeed, Michal, Philip, Lars, Caroline, Linda, Galin, Xiao, Madlen and Xiaoxiao; you all made my time at the chair enjoyable.

Tack so mycket to the folks at Combustion Physics, Lund University, Sweden. I appreciate you all very much. I must mention all these wonderful people I worked with: Prof. Marcus Alden, Prof. Per-Erik Bengtsson, Annelie, Cecilia, Gladys, Hadi, Terese, Harry, Per, Raffaella, Fikret, Adina, Kalle, Andreas, Aida, Karin and all the experimentalists. I am grateful to Dr. Martin Tuner for his improvements in the Stochastic Reactor Model. I wish to say thank you to my mentors Pastors Sunny Adeniyi, Mrs. Fola Adeniyi, Dele Bamgboye and Bishop David Oyedepo.

I give a lot of credit to my wife, Chioma for her patience and prayers and our children Asher and Charisa for their smiles and hugs. The financial support of the European Community project (LESSCO2) and Emissionbildung in Brennkammern von Fluggasturbinen' of the state of Brandenburg in Germany is gratefully acknowledged.

Ngozi Ebenezer
Fort St. John, BC, Canada
15. 05. 2014

ABSTRACT

In this work Adaptive Polynomial Tabulation (APT) is presented. It is a new approach to solve the initial value chemical rate equation system. In this approach zeroeth, first and second order polynomials are used in real-time to approximate the solution of the initial value chemical rate equation system. The sizes of the local regions encountered for the different orders of polynomial approximation are calculated in real-time. To improve accuracy the chemical state space is partitioned into hypercubes. During calculations the hypercubes accessed by the reactive mixture are divided into adaptive hypercubes depending on the accuracy of the local solution. Mixture initial conditions are stored in the adaptive hypercubes. Around each stored initial condition two concentric ellipsoids of accuracy (EOA) are defined. These include the ISAT and *identical* EOAs. The time evolution of mixture initial conditions which encounter an *identical* and ISAT EOA are approximated by zeroth and first order polynomials respectively. With a certain number of stored initial conditions within an adaptive hypercube, its second order polynomial coefficients are constructed from the stored initial conditions. The time evolution of additional mixture initial conditions that encounter this adaptive hypercube are approximated with second order polynomials. The APT model is simplified by the replacement of the entire set of species mass fractions with a progress variable based on the enthalpy of formation evaluated at 298 K. APT has 3 degrees of freedom which include the progress variable, total enthalpy and pressure. The APT model was tested with a zero dimensional Stochastic Reactor Model (SRM) for HCCI engine combustion. A skeletal n-heptane/toluene mechanism with 148 chemical species and 1281 reactions was used. In the tests, the HCCI engine simulations using APT were in very good agreement with the model calculations using the ODE solver. The cool flame and main ignition events were accurately captured. The major and minor species were also accurately captured by APT. In SRM-HCCI calculations without cyclic variations, a computational speed up factor greater than 1000 was obtained when APT was used for all the operating points considered without significant loss in accuracy. For the SRM-HCCI engine calculations with cyclic variations, APT demonstrated a computational speed up exceeding 12 without significant loss in accuracy.

Keywords: ISAT, PRISM, APT and SRM

Chapter 1   Introduction

1.1   Brief Background

Energy from fossil fuels contributes at least 87 % of the global energy demands. This percentage is likely to increase in the near future because of increasing industrialization of China and India. Although significant efforts have been put to develop alternative forms of energy such as hydrogen fuel cells, wind energy, bio-fuels, and solar energy, there are several challenges limiting their successful and sustainable implementation. For example hydrogen fuel cells offer zero emission hazards but they pose serious transportation risks. Bio-fuels will take a reasonable chunk of global food supply and wind energy sources are powered by energy and they require large areas of land. Our transportation system is powered by engines fueled by light and heavy hydrocarbons. The performances of these engines (SI, CI and HCCI) are far from optimum due to the complexity of combustion. Experimental testing and numerical simulations have been used extensively to study and understand engine combustion. Experimental testing methods include Laser Induced Fluorescence, Laser Induced Incandescence, Particle Image Velocity and Laser Doppler Anemometry. These methods have been used in studying hotspots and engine knock in SI engines, auto-ignition processes in HCCI engines, soot and NOx formation in Diesel (CI) engines. They require sophisticated and expensive equipment set up. Numerical combustion involves the coupling of computational fluid dynamics and computational chemical kinetics. This offers a method for validating experimental measurements and can provide results in situations where experimental measurements are prohibitive.

It is computationally expensive to incorporate the detailed chemistry of complex fuels (such as n-decane) into three dimensional computational fluid dynamics codes. This is because such mechanisms are large in size, and may contain hundred chemical species and several thousands of elementary reactions. The chemical species in the mechanism usually span a broad range of timescales. In engine environments, the temperature/species mass fractions local inhomogeneities have to be considered.

Several approaches have been used to simplify complex mechanisms and/or reduce the computational cost in solving the chemical rate equation system. These include convention reduction methods (QSSA [1], RCCE [2], chemical lumping [3]); dimension reduction methods (ILDM [4], CSP [5], ICE-PIC [6]) and storage/retrieval methods (ISAT [7], PRISM [8], neural networks [9], transient flamelet libraries [10]). Some of these methods can be used together.

The contribution of this work is based on a special class of storage/retrieval methods named solution mapping. Solution mapping involves approximation of the solution of the chemical ODE system in real time with simple algebraic polynomial expressions which are computationally cheaper to evaluate. PRISM and ISAT belong to this class. In simulating turbulent flames, PRISM and ISAT demonstrated significant computational speed up without significant loss in accuracy. These methods have scarcely been use for mechanisms with more than 50 chemical species. Particularly, PRISM has only been tested with hydrogen mechanism (9 species and 38 elementary reactions). As mentioned in [8] the memory requirements for PRISM is overwhelming, some local region coefficients were stored in an external disc file. However, ISAT Ellipsoids of Accuracy are smaller than the hypercubes in PRISM, therefore ISAT require more memory and searching time than PRISM. Memory requirements for ISAT and PRISM will be prohibitively large for larger chemical mechanisms.

In this work, first order (ISAT) and second order (PRISM) polynomial approximations to the solution of the chemical ODE system are combined into a single computer code named Adaptive Polynomial Tabulation [11]. It is anticipated that with 200 ISAT Ellipsoids of Accuracy in a PRISM (adaptive) hypercube, there will be a factor of 50 reductions in memory requirements for APT as compared to ISAT. In APT the entire set of species mass fractions is replaced by a monotonic reaction progress variable. This reduces sharply its memory requirements, enhancing its application to chemical mechanisms with hundreds of chemical species and thousands of elementary reactions. This merit of APT will be demonstrated in this thesis. Current real-time tabulated chemistry methods yielded success for flame calculations, but for engine applications during the expansion stroke the total enthalpy decreases because of volume work and heat loses. Therefore, there will be no initial conditions for the table entries during the engine expansion phase. In this work, with the improvement

3

associated with APT an attempt is made to resolve this problem. APT is tested in an engine environment providing the first opportunity to use real time second degree polynomials as a surrogate chemical ODE solver for engine calculations.

1.2   Problem to be solved

In this work, the successes and limitations of two solution mapping methods (In Situ Adaptive Tabulation-ISAT [7] and Piecewise Reusable Implementation of Solution Mapping-PRISM [8]) were considered and new method was built around them. This new approach possesses the best features of these two methods and an algorithm to cater for their limitations. ISAT uses first order polynomial expressions in real-time as approximations to the ODE solution. It possesses adaptive control of tabulation errors. Memory requirements increase quadratically with the number of species in the mechanism and the binary tree search for a nearest neighbour does not always give the nearest neighbour. In using this method with a perfectly stirred reactor (PSR) model [7], a computational gain factor 1000 was obtained.

PRISM approximates the ODE solution with second order polynomial expressions for each (hypercube) local region of the chemical composition space. Second order polynomial coefficients are constructed by calling the ODE integration solver at carefully selected initial conditions within the hypercube based on central composite design [12]. Polynomial

4

coefficients for each hypercube are calculated when the mixture initial conditions enters it for the first time. PRISM lacks adaptive control of tabulation errors and some coefficients are constructed but rarely used. The memory requirement for this method is high. The frequently used hypercube coefficients are stored in main memory while less frequently used ones are stored in an external disc file. The cost constructing second order polynomial is about 250 times that of one ODE integration call.

Adaptive Polynomial Tabulation (APT) combines ideas in PRISM and ISAT. It uses zeroeth, first and second order polynomials in real-time as approximation to the ODE solution. In APT the entire set of chemical species mass fractions is replaced by a progress variable, thereby drastically reducing the size of its storage/retrieval table. Adaptive local regions for different polynomial approximations are calculated on the fly from the mapping gradients. Contrary to PRISM, only stored initial conditions are used for the construction of second degree polynomial coefficients (the computational cost of constructing one second order polynomial in APT is far cheaper than one ODE integration call). If an initial condition encounters an ISAT Ellipsoid of Accuracy (EOA) of a stored initial condition, its ODE solution is approximated by either a zeroeth order or a first order polynomial. Initial conditions are stored outside the ISAT EOA of stored initial conditions. This tendency and the growth of the ISAT EOA enhance the spread of initial conditions within a PRISM hypercube and the accuracy of second order polynomial coefficients. Therefore, as initial conditions are

5

stored in a PRISM hypercube, APT gets zeroeth order/first order polynomials reuse improving its overall computational efficiency. The problem encountered in PRISM with second degree polynomial coefficients with limited or no reuse is completed avoided by APT. Before second order polynomials are constructed for a given PRISM hypercube enough zeroeth and first order polynomial reuses should have been recorded, second order polynomial coefficients are constructed only for PRISM hypercubes with N_p stored initial conditions. In APT, second order polynomial coefficients are constructed only as needed, and not because the reaction trajectory passes through the PRISM hypercube as implemented in the original PRISM formulation [8].

1.3 Solution approach

In APT, the chemical composition space is divided at the pre-processing into equally-sized hypercubes. APT computes with direct integration the ODE solution for the first initial condition that enters a given hypercube. The mapping gradient for this initial condition is also calculated. It is calculated by using 3 extra ODE integration calculations at small perturbations of the progress variable, total enthalpy and total pressure. The mapping gradient is used to compute the ISAT EOA and the PRISM hypercube size. As the calculation proceeds, initial conditions are stored in the PRISM hypercube, outside the ISAT EOA of previously stored initial conditions ensuring the spread of initial conditions within the PRISM hypercube. Although this spread of initial conditions cannot be compared to that of central composite design, but at least it gives a level of spread of initial conditions on its own right. As more initial conditions encounter this PRISM hypercube, those that fall within the ISAT EOA of a stored initial condition are approximated with either zeroeth order or first order polynomials. When the number of stored initial conditions for a given hypercube equals N_p, the stored initial condition information is used to construct second order polynomials for the PRISM hypercube. This approach is computationally cheaper approach than that used in the original PRISM formulation [8]. Initial conditions that encounter this PRISM hypercube will have their ODE solution approximated by second order polynomial coefficients. Therefore, APT provides zeroeth, first and second order polynomials in real-time as solution to chemical ODE system.

In order to test APT, a PDF-based testing tool was used because it is one of the accurate models that can capture turbulent-chemistry interaction in an engine environment. However, it is computationally expensive incorporate PDFs with 3 dimensional CFD code. As a first step in APT model development, the Stochastic Reactor Model that has been developed recently [13-16] was used instead of a full three dimensional CFD. The SRM involves similar processes such as turbulent mixing, pressure variation, chemical reaction, time marching and convective heat transfer as the complete CFD problem, but with a reduced computational cost. In this

7

thesis the SRM tests were limited to 100 computational particles, because reduced computational demands favors model development.

HCCI engines are known to be kinetically controlled; therefore their ignition timing will be sensitive to errors introduced by Adaptive Polynomial Tabulation. Recently, n-heptane/toluene blends were introduced as realistic reference fuels for HCCI engines with special low temperature combustion characteristics [17-18]. Mauss and co-workers recently developed skeleton kinetic models for this blend which was reduced with the help of linear lumping and species removal [3, 19-20]. The resulting skeletal mechanism contains 148 species and 1281 elementary reactions and it was used to simulate the sensitive HCCI experiments in reference [17]. For multi-cycle engine simulations with and without cyclic variations computational speed up exceeding 12 and 1000 were obtained respectively [11]. Finally, a reduced library APT model was proposed. In this case at most 15 initial conditions are stored per time step for the first 5 engine cycles for multi-cycle simulations with cyclic variations. After the fifth cycle, initial conditions are added if the cumulative number of stored initial conditions is less than 15 for each time step. The benefits of these simplifications are spread of initial conditions within each local region and a reduced size APT library. This version of APT was tested with the n-heptane/toluene fueled stochastic reactor model for HCCI engines. A computational speed up of 16 was obtained without significant loss in accuracy.

1.4 Summary of the work

This work pivots around the development of a new solution mapping method (APT) and its application to Stochastic Reactor Models for internal combustion engines [11]. In Chapter 2, solution mapping methods are described. In particular, PRISM and ISAT are explained. APT and the testing tool – SRM for HCCI engines are also described in Chapter 2. In Chapter 3, the calculations and results are presented for the test of APT with n-heptane/toluene fueled Stochastic Reactor Model for HCCI engines. Results for multi-cycle simulations with and without cyclic variations are also presented. The details of the accuracy of APT, tabulation errors associated with APT, local polynomial reuse, and computational speed up are discussed. The conclusion is presented in Chapter 4.

Chapter 2   Solution Mapping Methods and Their Applications

A special class of storage/retrieval methods known as solution mapping is introduced in this chapter. This model simplification targets the solution of the chemical rate equation system. Solution mapping removes the stiffness of the system and evaluates algebraic polynomials in real-time as approximations to the time demanding ODE solutions. These methods can be used in combination with other reduction methods. Examples include ISAT [7], PRISM [8] and APT [11].

2.1   In Situ Adaptive Tabulation (ISAT)

In this method, the solution of the initial value chemical rate equation system is approximated by first order polynomial expressions. These polynomials are constructed in real-time from the stored initial conditions in a look-up table. The chemical rate equations system is given by Equation (2.1).

$$\frac{\partial \phi}{\partial t} = S(\phi)$$
$$\phi_0 = \phi(t_0) \tag{2.1}$$

The chemical ODE system is composed of several elementary reactions whose rate constants may vary by many orders of magnitude. The chemical source term is nonlinear. An Ordinary Differential Equation (ODE) solver is used to solve Equation (2.1) numerically. It is computationally expensive to use the ODE solver for combustion simulations of practical interest. ISAT is one of the methods used as an alternative to this problem. In the ISAT procedure, for each initial condition ϕ_0, the sensitivity with respect to the initial condition or mapping gradient $A(\phi_0)$ is calculated using the following formula:

$$\frac{dA}{dt} = JA$$
$$A_0 = I \tag{2.2}$$

Where J is the Jacobian matrix and I is the Identity matrix. The Jacobian matrix is given by:

$$J = \frac{\partial S}{\partial \phi} \tag{2.3}$$

In some codes, Equations (2.1) and (2.2) are solved simultaneously. The ODE solution at ϕ_0 after time interval Δt is ϕ_0^{out}. If there are no stored records in the look-up table, then the ODE solver is called to compute ϕ_0^{out} and $A(\phi_0)$. The mapping gradient is used to calculate the semi-principal axes $\delta\phi_0^{l}$ of an ellipsoid centred at ϕ_0^{out}. This ellipsoid is known as the Ellipsoid of Accuracy (EOA).

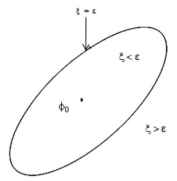

$\xi = \varepsilon$

$\xi < \varepsilon$

ϕ_0

$\xi > \varepsilon$

Figure 2.1: A sketch of the ISAT Ellipsoid of Accuracy (EOA) about the initial condition ϕ_0.

The length of the semi-principal axes $\delta\phi_0^I$ of the ellipsoid of accuracy can be calculated from the following formula:

$$A(\phi_0)\delta\phi_0^I = \varepsilon\phi_0^{out} \tag{2.4}$$

Suppose the ODE solution at an additional initial condition ϕ_1 is required from the ISAT method. If ϕ_1 is within the EOA of ϕ_0, then, the ODE solution at ϕ_1, is approximated by first order polynomials using Equation (2.5). This is referred to as retrieval.

$$\phi_1^{out} = \phi_0^{out} + A(\phi_0)(\phi_1 - \phi_0) \tag{2.5}$$

If ϕ_1 is outside the EOA of ϕ_0, the ODE solution at ϕ_1 and the mapping gradient $A(\phi_1)$ are computed by direct integration using the ODE solver. The error ξ in the first order polynomial approximation is calculated. If $\xi \leq \varepsilon$, then the new query initial condition ϕ_1 is not stored and the EOA of ϕ_0 is increased in size. The new EOA will contain the old EOA and the query initial condition ϕ_1. This is known as growth. It is shown in Figure 2.2.

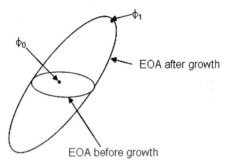

Figure 2.2: A sketch that shows the growth of an Ellipsoid of Accuracy (EOA). The new EOA contains the original EOA and the query initial condition ϕ_1.

If $\xi > \varepsilon$, then the first leaf becomes a node. A cutting plane is constructed between the stored initial condition ϕ_0 and the query initial condition ϕ_1. The cutting plane is defined by a vector v and a scalar γ, which are given by Equation (2.6) and Equation (2.7) respectively.

$$v = \phi_1 - \phi_0 \tag{2.6}$$

$$\gamma = v^{\mathrm{T}} \frac{(\phi_1 + \phi_0)}{2} \tag{2.7}$$

The query initial condition ϕ_q, its ODE solution ϕ_q^{out} and its mapping gradient are stored as a new record in the look-up table. This is known as addition. For any additional query initial condition ϕ, if $v^T \phi \le \gamma$, then ϕ lies closer to ϕ_0 and if $v^T \phi > \gamma$, then ϕ lies closer to ϕ_1. Before growth, addition or retrieval can be accomplished, the ISAT procedure determines the closest stored initial condition to the query initial condition. This is performed with the binary search tree algorithm. The search starts from the top node, using its cutting vector v and cutting scalar γ. In this method the searching time for the closest stored record to a given initial condition is directly proportional to $\mathrm{Log}_2 n$, where n is the number of records in the look-up table.

2.2 Piece-wise Reusable Implementation of Solution Mapping (PRISM)

In this method, the solution of the initial value chemical rate equation system is approximated locally with second order polynomials. The chemical rate equation system is given by Equation (2.1). The PRISM method presents a computationally cheaper alternative to the direct numerical integration of Equation (2.1). It partitions the space of state variables a priori into block-shaped structures or hypercubes as shown in Figure 2.4. The state space spans all possible chemical species mass fractions, total enthalpy, and pressure. The hypercubes in PRISM are equally sized. Each hypercube possesses a unique integer index. The ODE solver receives an initial condition ϕ_0 evolves it over a time interval Δt to give a new set of species mass fractions ϕ_0^{out}.

Memory for each hypercube is created only when the reactive mixture enters it for the first time. During calculations, if a mixture initial condition ϕ_0 encounters a hypercube for which the second order polynomial coefficients have not been constructed, the PRISM algorithm samples this hypercube, creates memory for the hypercube and constructs its second order polynomial coefficients. It also evaluates the polynomials at the initial condition ϕ_0 to give a new set of species mass fractions ϕ_0^{out}. The coefficients are constructed from a generated set of initial conditions using central composite design [12]. In this design, $2^{k-p'} + 2k + 1$ initial conditions are used to construct the second order polynomials. There are $2^{k-p'}$ initial conditions at the edges of the hypercube from the fractional factorial design, $2k$ (star) initial conditions located outside each of the faces of the hypercube and 1 (centre) initial condition at the centre of the hypercube. This is shown in Figure 2.5. The star initial conditions serve for smoothing of the second order polynomials and the centre initial condition is used for checking the curvature of the second order polynomials. Central composite design gives

15

a diagonal covariance matrix. The $2^{k-p'} + 2k + 1$ initial condition and their

ODE solutions are used to construct $\left(1 + k + \dfrac{k(k+1)}{2}\right) n_s$ polynomial

coefficients using linear least squares.

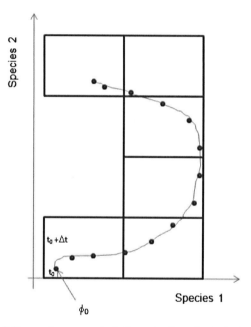

Figure 2.4: Temporal progress of a 2-dimensional reaction trajectory.

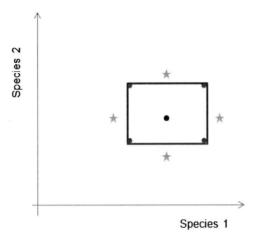

Figure 2.5: A sketch of the generated initial conditions used to construct second order polynomials. The circles denote the edge initial conditions. The stars denote the star initial conditions and the black circle at the centre of the hypercube indicates the centre initial condition. This is a two dimensional sketch of a hypercube.

The polynomial coefficients computed for each hypercube are stored either in the memory or in a disc file. Suppose the ODE solution at an additional initial condition ϕ_i is required, a tree search is performed to determine the index of the hypercube that contains ϕ_i. If second order polynomial coefficients for this hypercube have been constructed, then they are retrieved and evaluated at ϕ_i using Equation (2.8).

$$\phi_{1,i}^{out} = a_{i,0} + \sum_{j=1}^{n_s+2} a_{i,j}\phi_{1,j} + a_{i,n_s+3}\Delta t + \sum_{j}^{n_s+2} \sum_{k \leq j}^{n_s+2} a_{i,jk}\phi_{1,j}\phi_{1,k} +$$

$$a_{i,(n_s+3)(n_s+3)}\Delta t\Delta t + \quad \text{cross terms}$$

(2.8)

If this hypercube does not possess polynomial coefficients, then the polynomial coefficients are constructed as mentioned previously.

2.3 Adaptive Polynomial Tabulation

APT was developed from ISAT [7] and PRISM [8]. ISAT and PRISM approximate the solution of the chemical rate equation system in real time with first order and second order polynomials respectively. The memory requirement for these methods is high. Due to high memory requirements, PRISM has been limited to simulations involving small chemical mechanisms such as that of Hydrogen. It was used to simulate Hydrogen turbulent flame employing a chemical mechanism with 9 chemical species and 38 elementary reactions [8]. PRISM's hypercubes are of the same size (non-adaptive), therefore their polynomial coefficients may become inaccurate when the reacting mixture is progressing through regions in the chemical composition space where the state variables are changing fast. Hypercube coefficients in PRISM are constructed using extra ODE integration calculations at carefully selected points within the hypercube through central composite design [12]. The extra ODE integration cost must be recouped first before PRISM can record computational speed-up. In ISAT, the Ellipsoids of Accuracy drawn around each stored initial condition is adaptive, the size of the EOAs are calculated at runtime from the mapping gradients. ISAT uses binary search trees to determine the closest record to a given initial condition. Binary tree searches are fast and their search duration is directly proportional to $\text{Log}_2 n$. Although for large values n searching times might not be small. ISAT Ellipsoids of Accuracy are smaller than PRISM hypercubes, therefore the ISAT method requires more memory and longer searching time than the PRISM method. Now combining ISAT and PRISM in one computer code (APT) and if 200 initial conditions are stored in a PRISM hypercube, it is expected that memory requirements for the storage/retrieval table will decrease by a factor 50. The entire set of species mass fractions is replaced by a monotonic progress variable. A progress variable based on the enthalpy of formation evaluated at 298 K was successfully used to generate transient flamelet libraries in a Diesel combustion application [10]. In this thesis, this progress variable was adopted and it is defined as:

$$c(t) = \frac{H_{298}(t) - H_{298}(0)}{H_{298}(t_\infty) - H_{298}(0)} \tag{2.9}$$

This approach drastically reduces the memory size of APT's storage/retrieval table. APT provides zeroeth, first and second order polynomials as solutions to the chemical ODE equation system. With these changes APT can be used to simulate the combustion of higher hydrocarbon fuels. To accommodate situations where pressure is transient, such as in engines, the mapping gradient is calculated by employing small perturbations in the axes of progress variable, total enthalpy and pressure. Therefore, each mapping gradient calculation involves 3 extra ODE

19

integration calculations. During calculations initial conditions are stored in the PRISM hypercube. They are stored outside the ISAT EOAs of stored initial conditions. The ODE solutions for those initial conditions that encounter EOAs of stored initial conditions are approximated by zeroeth or first order polynomials. The problem encountered in the original PRISM formulation where some hypercube coefficients were constructed with few or no reuse is completed avoided in APT. When the reacting mixture passes through a hypercube, zeroeth order and first order polynomial reuse are first recorded boosting APT's computational speed up before the algorithm decides on constructing second order polynomials. In APT initial conditions for each PRISM hypercube are stored outside the ISAT EOAs enhancing their spread within the PRISM hypercube. This also improves the accuracy of its second order polynomial coefficients. In the pre-processing stage of calculations, the space spanned by progress variable, total enthalpy and pressure is divided into block-shaped structures or hypercubes of the same size as shown in Figure 2.6. Logarithmic scale was chosen for the axes of progress variable and pressure and a linear one for total enthalpy. Each hypercube is assigned a unique index. Their boundaries were carefully assigned. Accessing each hypercube's stored data is facilitated by its integer index. For example, for an initial condition ϕ_0, the index of the hypercube that contains ϕ_0 is easily determined using linear search tree function.

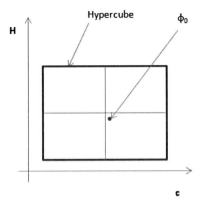

Figure 2.6: A two dimensional representation of the chemical composition space with the hypercubes and a query initial condition ϕ_0.

During calculations each hypercube may be divided into several PRISM hypercubes, when an initial condition ϕ_0 enters a given hypercube for the first time, its ODE solution is computed by direct integration. The mapping gradient is calculated by calling the ODE solver 3 times. From the mapping gradient, the size of the ISAT EOA is calculated using Equation (2.4). Two additional EOAs namely *identical* EOA and PRISM EOA are calculated from the ISAT EOA. They are calculated using the following formulae:

$$\delta\phi_0^0 = \beta_0 \delta\phi_0^1, \tag{2.10}$$

$$\delta\phi_0^2 = \beta_2 \delta\phi_0^1, \tag{2.11}$$

In this thesis, 1/1000 and 120 was used as values for β_0 and β_2 respectively. Therefore the PRISM EOA is a multiple of the smallest ISAT EOA while *identical* EOA is a fraction of the ISAT EOA. The PRISM EOA encompasses a PRISM hypercube as shown in Figure 2.7. Since it is the first initial condition that encountered this hypercube, the size for first PRISM hypercube is initialized with the hypercube size. The current PRISM hypercube size is updated depending on the number of division it performs. That is, as long as the product of β_2 and ISAT EOA size is less than the current PRISM hypercube size, it is divided into two equal halves and new leaves are added to the binary tree that organizes its data. This is represented in Figure 2.8. This process continues until product of β_2 and ISAT EOA size becomes greater than the current PRISM hypercube size. After this process, this initial condition-slope data is stored in the data structure for this PRISM hypercube. If a new initial condition ϕ_j enters this PRISM hypercube and outside the ISAT EOA of ϕ_i, the ODE solution for ϕ_j will be calculated by ODE integration. Before storing ϕ_j, the error (ξ) in the first order polynomial approximation to its ODE solution is calculated. If $\xi \leq \varepsilon$, then the EOA at ϕ_i is increased in size to contain ϕ_j and ϕ_j is not stored. This is known as growth and it is represented in Figure 2.2. On the contrary,

22

if $\xi > \varepsilon$ then, the ISAT EOA size and the mapping gradient are calculated. The current PRISM hypercube size will be divided repeatedly until it becomes less than the product of β_2 and ISAT EOA size.

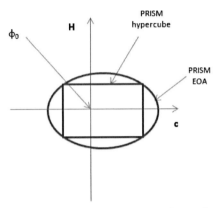

Figure 2.7: An initial condition within an adaptive hypercube. The adaptive hypercube is fully covered by the PRISM EOA. The adaptive hypercubes are used because they are easily accessed by the binary tree functions.

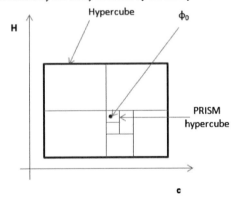

Figure 2.8: A two dimensional representation of the chemical composition space with the hypercubes, adaptive hypercubes and initial condition ϕ_0.

After each division, new leaves are added to the binary tree that organizes the data for the PRISM hypercube. Each PRISM hypercube has its own index or leaf, when it is divided it becomes a node. Each node has a cutting axis and a cutting value. For any given initial condition ϕ that enters this hypercube, a binary search tree function is used to determine the leaf or hypercube index of the PRISM hypercube that contains ϕ.

Hypercubes are easily accessed through linear search tree, and the PRISM hypercube within a hypercube are accessed using a binary search tree. Initial conditions within a PRISM hypercube are stored in a manner that facilitates easy and faster accessibility. Each initial condition-slope data has a leaf associated with it. If ϕ_1 is the first initial condition that enters a PRISM hypercube then it is assigned a leaf. If ϕ_1 is the second initial condition to be stored in this PRISM hypercube, then, the initial leaf is replaced by a node with two daughter leaves. The cutting plane of the node that divides ϕ_0 and ϕ_1 is given by:

$$v^T \phi = \gamma \tag{2.12}$$

Expressions for the cutting scalar and cutting vector are shown in Equations (2.6) and (2.7). Suppose a new initial condition ϕ_2 enters this PRISM hypercube, if $v^T \phi_2 \leq \gamma$ then, ϕ_2 lies very close to ϕ_0. If $v^T \phi_2 > \gamma$ then ϕ_2 lies close to ϕ_3. In either case a new node can be created with a cutting plane defined in terms of ϕ_2 and either ϕ_1 or ϕ_3. This is how initial conditions are stored in the binary tree that organizes the data for the PRISM hypercube.

Where the number of stored initial conditions for a given PRISM hypercube equals N_p, its second order polynomial coefficients are constructed. The stored initial conditions and their ODE solutions are used. The N_p stored initial conditions and ODE solutions are represented in matrix $\hat{B}$ and $\hat{Y}$ respectively. Principal Component Analysis [22] is employed to transform $\hat{B}$ into $\breve{B}$ and every redundancy in $\hat{B}$ is removed. A second order matrix $\tilde{B}$ is created from $\breve{B}$. The product of $\tilde{B}$ and $\hat{Z}$ gives the following over-determined system of equation:

$$\tilde{B}\hat{Z} = \hat{Y} \qquad\qquad (2.13)$$

Pre-multiplying both sides of Equation (2.13) by $\tilde{B}^T$ gives the following equation

$$\overline{B}\hat{Z} = \overline{Y} \qquad\qquad (2.14)$$

Equation (2.14) is solved using Lower Upper decomposition algorithm and back substitution [22].

Suppose the SRM asks APT for the ODE solution at an initial condition ϕ_q. The reduced initial condition ϕ_q^r is derived from ϕ_q. In order to

25

perform this task, APT starts by determining index of the hypercube that contains ϕ_q^r using a linear search tree. Within this hypercube, the index of the PRISM (adaptive) hypercube that contains ϕ_q^r is determined using binary search tree. APT later verifies whether this PRISM hypercube has second order polynomial coefficients. If it has second order polynomial coefficients and ϕ_q^r is within its allowed or α defined section, then, second order polynomials are retrieved and evaluated at ϕ_q^r using Equation (2.15). This is known as PRISM evaluation. The length of the allowed of a PRISM hypercube is the product of α and the standard deviation of the stored initial conditions.

$$\phi_{q,i}^{out} = a_{i,0} + \sum_{j=1}^{m'} a_{i,j} \hat{\phi}_{q,j}(t) + \sum_{j=1}^{m'} \sum_{k\geq j}^{m'} a_{i,jk} \hat{\phi}_{q,j}(t) \hat{\phi}_{q,k}(t) \tag{2.15}$$
$$i = 1, 2, ..., n_s$$

PCA [21] was used in the determination of the second order polynomial coefficients. PCA eliminates the influence of redundant variables. If there are no second order polynomial coefficients, APT checks whether this PRISM hypercube has stored initial conditions. A binary search is performed to determine the closest stored initial condition ϕ_q^r. The stored initial conditions, *identical* EOA and ISAT EOA for a PRISM hypercube are shown in Figure 2.9. APT test whether ϕ_q^r is within the ISAT EOA of ϕ_q^r. If this test holds, a new test is performed to verify if ϕ_q^r is within the

26

identical EOA of ϕ_s^r. The query initial condition is taken as the stored initial condition if ϕ_q^r is within the *identical* EOA of ϕ_s^r. This is known as *identical* evaluation. If the previous test was false, the ODE solution at ϕ_q^r is approximated using Equation (2.16). This is known as ISAT evaluation.

$$\phi_q^{out} = \phi_s^{out} + A(\phi_s)(\phi_q^r - \phi_s^r) \qquad (2.16)$$

In case there is no stored initial condition in this PRISM hypercube or ϕ_q^r is outside the ISAT EOA of ϕ_s^r, the ODE solution at ϕ_q^r is calculation using ODE integration. The error (ξ) in the first order polynomial approximation to the ODE solution at ϕ_q^r is calculated if ϕ_q^r is outside the ISAT EOA of ϕ_s^r. If $\xi \leq \varepsilon$, then the ISAT EOA at ϕ_s^r is increased in size to contain ϕ_q^r. This is known as growth. The query initial condition is not stored in this case. If the error is greater than the ISAT error tolerance or ϕ_q^r is the first initial condition in the PRISM hypercube, the mapping gradient and ISAT EOA size are calculated. The current PRISM hypercube is divided repeatedly into two equal halves until it is smaller than the calculated PRISM hypercube size. During each division, new leaves are added to the binary tree structure that organizes the data for the PRISM hypercube. Finally, if the number of stored initial conditions equals N_p, then the stored initial conditions and their ODE solutions are used to construct second order polynomials for the

27

PRISM hypercube. The second order polynomial coefficients are stored in the data structure for the PRISM hypercube and the ODE solution computed by ODE integration is returned to the SRM. This is shown in Figure 2.10.

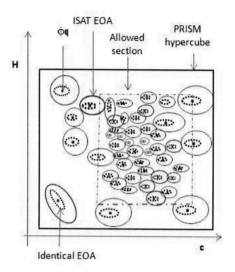

Figure 2.9: A PRISM hypercube with stored initial conditions is shown. Each initial condition is encircled with an *identical* EOA and an ISAT EOA. The PRISM hypercube is divided into two sections; an allowed section and disallowed section.

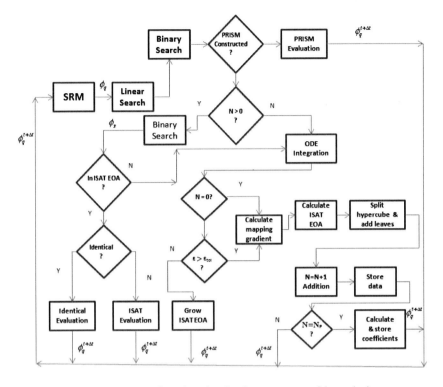

Figure 2.10: An APT flow chart showing the components of the method

2.3.1 Reduced library APT model

APT uses zeroeth, first and second order polynomials as approximations to the ODE solution. The second order polynomials are constructed from stored initial conditions rather than selected initial conditions determined by central composite design. Central composite design [12] was used for the construction of second order polynomial coefficients in the original PRISM formulation [8]. It ensures the spread of initial conditions within a hypercube, initial conditions are located mostly at the edges and some outside each face. The spread of initial conditions increases the accuracy of the polynomials.

When APT is called within the stochastic reactor model, a situation can occur when all the initial conditions in a PRISM hypercube have the same pressure (no spread along the pressure axis). This decreases the accuracy of the second order polynomial approximation. To circumvent for this problem, a reduced library APT model was proposed. In this case at most 15 initial conditions are stored per time step for each of the first five cycles. After the fifth cycle, a new initial condition can be added if the cumulative sum of stored initial conditions after the fifth cycle is less than 15 per time step. This approach has three benefits. Firstly, it gives a reduced size APT library. Secondly, it improves the spread of initial conditions within the hypercube particularly along the pressure axis and hence improvement in the accuracy of the second order polynomial coefficients. Thirdly, it avoids the redundant time demanding ODE integration calculations for the construction of the mapping gradients.

In the next section, the SRM for HCCI engine is described. This is the testing tool for APT in this thesis.

2.4 Stochastic Reactor Model for HCCI engines.

HCCI engines inhale a uniform gas mixture in their combustion chambers and ignition occurs through compression. These engines are considered as future alternatives to Diesel and Spark Ignition engines. Their merits include are low soot and NOx emissions and high fuel efficiency at part load conditions. HCCI engines mostly operate with lean fuel mixtures and self-ignition occurs at the same time at several locations inside the combustion chamber. The lean fuel/air mixture burns at a lower temperature yielding low heat loss, high fuel economy, less NOx and soot are produced. Both High pressure injection and throttling are not required. HCCI engines yield better engine performance at lower cost. However, there are challenges associated with their successful utilization. The combustion process in HCCI engines is hard to control. Additional problems associated with HCCI engine combustion are the large emissions of unburned hydrocarbons and carbon monoxide due to the low temperature and lean combustion process. They can encounter combustion instability near stoichiometric mixtures giving rise to high pressure fluctuations which may damage them.

In the SRM the gas mixture in the engine cylinder is modeled as a single zone stochastic reactor. It considers scalars such as temperature, chemical species mass fraction, density, progress variable as random variables with a certain probability distribution. The PDF transport equations for these scalars are derived using the statistical homogeneity assumption. The progress variable has been included as random variable in this thesis. Also quantities such as total mass, volume, mean density and pressure are considered as global quantities. The global quantities do not vary spatially within the combustion chamber. The chemical species mass fractions Y_i, temperature T and progress variable c, vary within the combustion chamber and these random variables can be expressed as $\phi(t) = (Y_1(t), Y_2(t), ..., Y_{n_s}(t), T(t), c(t))$. For variable density flows the SRM is represented in terms of the Mass Density Function (MDF). The corresponding MDF is represented by $F_\phi(\varphi_1, \varphi_2, ..., \varphi_{n_s+2} ; t)$. The Equation (2.17) represents the time evolution of the MDF.

$$
\frac{\partial F_\phi(\varphi;t)}{\partial t} + \frac{\partial}{\partial \varphi_i}\left(Q_i(\varphi)F_\phi(\varphi;t)\right) + \frac{\partial}{\partial \varphi_{n_s+1}}\left(U(\varphi_{n_s+1})F_\phi(\varphi;t)\right)
$$
$$
= \frac{C_\phi \beta_c}{\tau} \int_{\varphi_j} \int_{\varphi_k} \delta(\varphi - \frac{1}{2}(\varphi_j + \varphi_k)) F_\phi(\varphi_j) F_\phi(\varphi_k) d\varphi_j d\varphi_k \quad ; j \neq k
$$
$$
F_\phi(\varphi;0) = F_\phi^0(\varphi)
$$

(2.17)

The right hand side of Equation (2.17) gives the effect of mixing on the MDF. The mixing is performed with the Curl mixing model. In this model the mixing takes place in randomly selected particle pairs. It is relatively simple to use and has good performance. The terms Q_i denote the change of the MDF due to chemical reactions and change in volume. The ODEs for species reaction rates and temperature are solved deterministically using a

Backward Differentiation Formula (BDF) method of order 5 or Adaptive Polynomial Tabulation.

$$Q_i = \frac{M_i}{\rho}\,\omega_i \quad i = 1, 2, ..., n_s \tag{2.18}$$

$$Q_{n_s+1} = p\frac{1}{mc_v}\frac{dV}{dt} + \frac{1}{c_v}\sum_{i=1}^{n_s}\left[\left(h_i - \frac{RT}{M_i}\right)\frac{M_i}{\rho}\,\omega_i\right] \tag{2.19}$$

$$Q_{n_s+2} = \frac{M_i}{\rho}\,\omega_{n_s+2} \tag{2.20}$$

The third term on the left hand side of Equation (2.17) is the convective heat loss term.

$$U = \frac{-h_g A''}{mc_v}(T - T_w) \tag{2.21}$$

To introduce the fluctuations, the convective heat loss term is replaced by the finite difference scheme (Equations (2.22) and (2.23)).

$$\frac{1}{h'}\Big(U(\varphi_{n_s+1})F(\varphi,t)\Big) \;-$$
$$\frac{1}{h'}\Big(U(\varphi_{s+1} - h')F(\varphi,..,\varphi_{n_s+1} - h',t)\Big), \;\; U(\varphi_{n_s+1}) < 0 \tag{2.22}$$

$$\frac{1}{h'}\Big(U(\varphi_{n_s+1})F(\varphi,t)\Big) \; - $$

$$\frac{1}{h'}\Big(U(\varphi_{s+1}+h')F(\varphi_1,..,\varphi_{n_s+1}+h',t)\Big), \; U(\varphi_{n_s+1}) > 0 \qquad (2.23)$$

The procedure for incorporating the convective heat transfer step follows the ideas presented in Ref [14-15]. The formulation is such that the stochastic model for convective heat loss in the limit approaches the deterministic Woschni correlation.

An equi-weighted Monte-Carlo particle method [13-16] with second order time splitting algorithm is employed to solve Equation (2.17) numerically. This method involves the approximation of the initial MDF by an ensemble of stochastic particles and the particles are moved according to the evolution of the MDF. Thus, depending on the internal Exhaust Gas Recirculation (EGR) mass ratio at Inlet Valve Closure (IVC) and the composition of the fresh air-fuel mixture, the SRM calculates the average initial mass fraction of the chemical species.

All the other stochastic particles in the ensemble are initialized with the fresh gas composition and temperature at IVC. In this thesis internal EGR was considered, some particles were introduced containing EGR only. The time t_0 corresponds to the time at IVC and Δt is the deterministic global time step used for the operator splitting. The time marching, convective heat loss, mixing and chemical reaction events are performed on the particle ensemble. The final time for this loop is at Exhaust Valve Opening (EVO) [14-15]. A pressure correction algorithm is included after each time step to

34

equalize the pressure of all the stochastic particles. This is because the thermodynamic properties of the particles change after each event causing their pressure to change. The pressure correction algorithm is explained in [16].

In the next chapter the calculations with SRM-HCCI engine model are presented for a large chemical mechanism involving more than 100 chemical species. The chemistry is solved either with APT or the ODE integration solver.

Chapter 3   Combustion Engine Calculations

3.1   N-heptane/toluene fueled SRM-HCCI calculations

In this section the results obtained from SRM-HCCI engine calculations that employed APT to perform its chemical reaction step is presented. A n-heptane/toluene mechanism with 148 chemical species and 1281 elementary reactions was used for the calculations. It was developed by Mauss and co-workers [3, 19-20] and validated by Kalghatgi and co-workers [17, 18] using shock tube experiments. The experiments were performed with a single cylinder engine based on SCANIA D12 as mentioned in [17]. The fuel injection took place at bottom dead centre with a port fuel injection system. Boost pressure was supplied by an external compressor and it could reach a maximum of 6 bar. A feedback control system ensured that the intake temperature was within 2 $^{\circ}$C of its chosen temperature. No external EGR was used.

3.1.1   Demonstration of the SRM-HCCI engine model

In this section the SRM-HCCI engine model is compared with engine experiments. These engine experiments were performed for different blends of n-heptane and toluene. Four operating points from the work of Kalghatgi et al [17-18] were considered. The engine settings are shown in Table 1. The cases selected are shown in Tables 2. Case 1 and Case 2 were run under the same engine settings but the fuel in Case 1 contains 65% of toluene while that of Case 2 contains 75% of toluene. The experimental measurements showed an earlier auto-ignition for Case 1 as compared to Case 2. This is because the aromatic hydrocarbon toluene is known to have a high octane number and it causes delay to auto-ignition. Case 4 was run with a lower inlet temperature and with a relatively high boost pressure. This case showed the earliest auto-ignition.

Table 1: Engine geometrical settings

Parameter	Value
Bore	0.127 m
Stroke	0.154 m
Rod	0. 255 m
Displaced Volume	1.95×10^{-3} m^3
Compression ratio	16.7
Inlet Valve Closure	-139 crank angle degree
Exhaust Valve Opening	121 crank angle degree

Table 2: Fuel and engine operating conditions

Cases	n-heptane	Toluene	Engine speed (rpm)	Intake temperature (^oC)	Intake pressure (bar)	Mixture Strength (λ)
1	35 %	65 %	900	120	1.0	3.5
2	25 %	75 %	900	120	1.0	3.5
3	25 %	75 %	1200	120	1.0	3.0
4	32 %	68 %	900	40	2.0	4.0

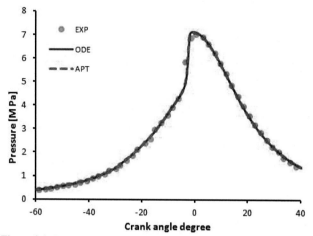

Figure 3.1: Pressure histories for Case 1. The experimental measurement is represented as a green circle, the ODE integration curve is represented as blue line and the APT curve is represented as a dashed red line.

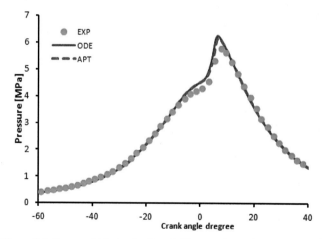

Figure 3.2: Pressure histories for Case 2. The experimental measurement is represented as a green circle, the ODE integration curve is represented as blue line and the APT curve is represented as a dashed red line.

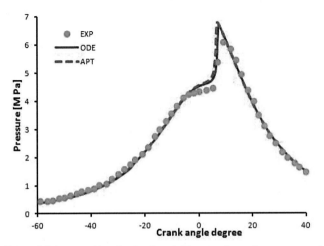

Figure 3.3: Pressure histories for Case 3. The experimental measurement is represented as a green circle, the ODE integration curve is represented as blue line and the APT curve is represented as a dashed red line.

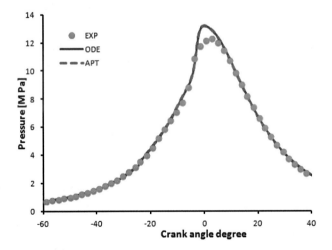

Figure 3.4: Pressure histories for Case 4. The experimental measurement is represented as a green circle, the ODE integration curve is represented as blue line and the APT curve is represented as a dashed red line.

As shown in Figures 3.1–3.4, the experimentally measured pressure histories are plotted on the same scale with those obtained from SRM-HCCI (APT and ODE integration) engine calculations. In each case APT and ODE integration accurately captures the experimentally measured pressure traces. It shows that these tests are under engine relevant conditions and the method can also be applied for more complex engine calculations. In Case 4 the SRM slightly over-predicts the experiment. This is due to the limitations of the Curl mixing model and the zero dimensional engine models used in this thesis. For Case 2 and Case 3 the main excitation event occurred after top dead centre and the heat release from the combustion competes with heat loss due to the cylinder expansion. Therefore, small changes in the ignition temperature can cause local quenching of chemical reactions because the chemical reactions rate terms are temperature dependent. Error can easily be introduced in the APT predictions because the sensitive tendencies of these cases.

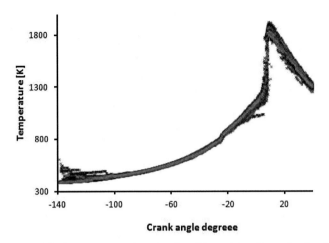

Figure 3.5: Temperature trajectories for 100 particles illustrating local inhomogeneities. SRM used ODE integration to calculate the chemical kinetics.

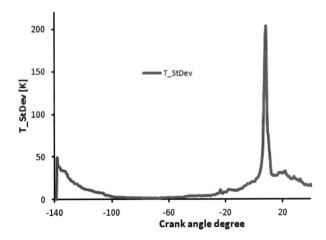

Figure 3.6: The variation of standard deviation of temperature with runtime in crank angle degrees. SRM used ODE integration to calculate the chemical kinetics.

The SRM considers a fairly homogeneous charge at IVC. Only particles containing internal EGR introduce the inhomogeneity. During calculation, a particle is randomly selected at each instant to perform the stochastic 'jump' step in temperature with respect to the wall temperature. This stochastic step depends on an exponentially distributed waiting time. This gives rise to realistic heat losses over a controlled number of particles. As shown in Figure 3.5, there is a spread in the ignition timing of about 5 CAD and some few particles with very large deviations. The variation of the standard deviation with runtime in crank angle degrees is shown in Figure 3.6. The peak standard deviation is about 210 K and it occurs at about 8 CAD during the main excitation of the blend.

In the next sections, two types of SRM calculations are presented. In one case, The SRM-HCCI engine is run with the same random seed (Repeated Single Cycle calculations) for all the cycles and ten engine cycles are considered for this case. In the second case, the SRM-HCCI engine is run with different random seed (Free Stochastic Cycle calculations). Thirty engine cycles are considered for FSC calculations and they show cyclic variations. These appear similar to cyclic variations of real engines. The SRM can predict the engines relative sensitivity on cyclic variations. However their amplitudes are strongly depending on the numerical accuracy of the SRM, that is, the number of particles or the time step size of the operator splitting method.

3.1.1.1 Demonstration with Repeated Single Cycle calculations

It is illustrated in this section that using APT HCCI engine auto-ignition is predicted with the same accuracy as the direct ODE integration. These results are obtained for different n-heptane/toluene blends and engine operating conditions.

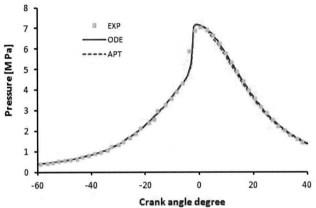

Figure 3.7: Pressure histories for Fuel D OP3 in RSC calculations. The mean pressure profiles of the ODE and APT calculations are shown in blue line and red dashed line respectively. The green symbol denotes the experimental measurements.

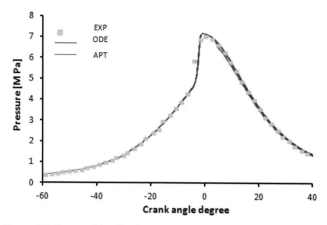

Figure 3.8: Pressure histories for Fuel D OP3 in RSC calculations. The 10 individual cycles with chemistry calculated by using ODE integration and the APT models are shown in blue lines and red lines respectively. The green square symbol denotes experimental measurements

In Figures 3.7 and 3.8, the experimental in-cylinder pressure is compared to the simulated SRM-HCCI engine model (APT and ODE integration) for Case 1. The chemistry is calculated by either ODE integration or the APT look up. In Figures 3.7, the mean pressure profile calculated using ODE integration is in good agreement with the mean pressure obtained using the APT model. In Figures 3.8, the pressure of the individual cycles obtained from the ODE integration calculations are in good agreement with those obtained from the APT calculations. The experimentally measured pressure profile is faithfully captured by the mean ODE integration and mean APT pressure curves as well the ODE integration and APT pressure profiles of the respective cycles as shown in

44

Figures 3.7 and 3.8. No cyclic variation is observed as shown in Figure 3.8 because the SRM is locked. It should be noted that for the first cycle the EGR composition is read from an external file and for subsequent cycles it is specified as evaluated from the SRM code.

3.1.1.2 Demonstration with Free Stochastic Cycle calculations

In Figures 3.9-3.16 the measured and computed in-cylinder pressure for the four engine cases are shown. The main ignition event for Case 1 and Case 4 occurs before the Top Dead Centre.   Ignition occurs during compression and differences in ignition temperature will result in only small differences in ignition timing. Therefore these cases are less sensitive to cyclic variations. As shown in Figure 3.9 and Figure 3.15 the region around the TDC is zoomed in order to emphasize the cyclic variations present. In these two cases there is very good agreement between the measured pressure and the calculated (APT and ODE integration) pressure curves. This is also shown for the mean pressures for Case 1 and Case 4 in Figures 3.10 and 3.16 respectively. In both cases SRM calculations using APT and ODE integrations are in very good agreement. The pressure profiles for 30 individual cycles for calculations that used APT and ODE integration are in very good agreement. However, there is a noticeable deviation between these models and experiment near the top dead centre for Case 4 as shown in Figures 3.15 and 3.16. This deviation is caused by assumptions in the zero dimensional SRM and limitations of the Curl mixing model. For Cases 2 and 3 as illustrated in Figures 3.11-3.14 the main ignition event occurs after the TDC. In these cases combustion is completed late in the expansion stroke. The pressure rise resulting from the chemical reactions is competing with the pressure reduction from the expanding cylinder volume. Small changes in the ignition temperature will result in remarkable changes in ignition timing. Thus these cases are very sensitive and show stronger cyclic variations in the engine. This in turn puts a high demand on the accuracy of the APT model. As shown in Figures 3.11 and 3.13, the pressure histories for the 30 individual engine cycles demonstrate significant cyclic variations close to the Top Dead Centre. The mean pressure profiles for SRM calculations using APT and ODE integration are in good agreement as shown in Figures 3.12 and 3.14 for Cases 2 and 3 respectively which demonstrates the accuracy of the progress variable and APT.

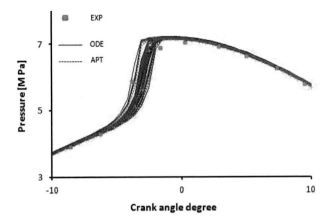

Figure 3.9: Pressure histories for Fuel D OP3 (Case 1) for Free Stochastic calculations. The 30 individual cycles with the chemistry calculated using ODE integration and APT are shown in blue and dashed red lines respectively. The green symbol represents the experimental measurements

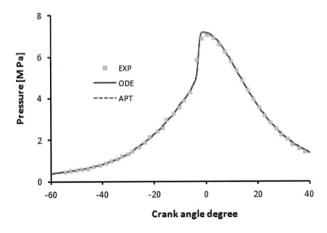

Figure 3.10: Pressure histories for Fuel D OP3 (Case 1) for Free Stochastic calculations. The mean pressure profiles for the SRM calculations using

ODE solver and APT models are shown with a blue line and red dashed line respectively. The green symbol represents the experimental measurements.

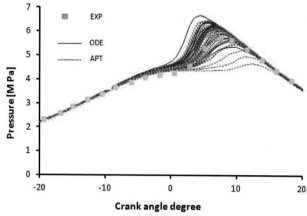

Figure 3.11: Pressure histories for Fuel C OP3 (Case 2) for Free Stochastic calculations. The 30 individual cycles with the chemistry calculated using ODE integration and APT are shown in blue and dashed red lines respectively. The green symbol represents the experimental measurements.

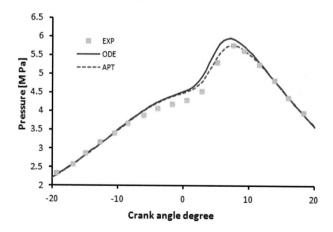

Figure 3.12: Pressure histories for Fuel C OP3 (Case 1) for Free Stochastic calculations. The mean pressure profiles for SRM calculations using the ODE integration and APT models are shown with a blue line and red

dashed line respectively. The green symbol represents the experimental measurements.

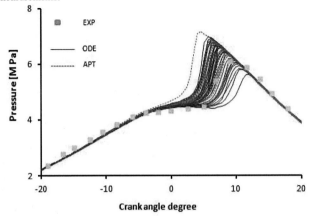

Figure 3.13: Pressure histories for Fuel C OP4 (Case 3) in the Free Stochastic calculations. The 30 individual cycles with chemistry calculated using ODE integration and APT are shown in blue and dashed red lines respectively. The green symbol represents the experimental measurements.

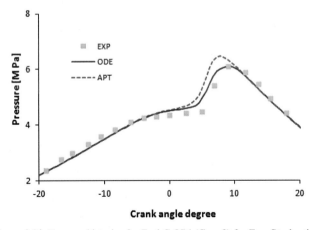

Figure 3.14: Pressure histories for Fuel C OP4 (Case 3) for Free Stochastic calculations. The mean pressure profiles for SRM calculations using ODE and APT models are shown with blue and red dashed lines respectively. The green symbol represents the experimental measurements.

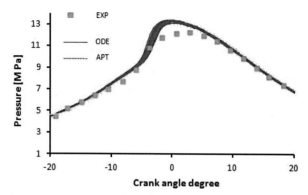

Figure 3.15: Pressure histories for Fuel E OP1 (Case 4) in the Free Stochastic calculations. The 30 individual cycles with chemistry calculated using ODE integration and APT are shown in blue and dashed red lines respectively. The green symbol represents the experimental measurements.

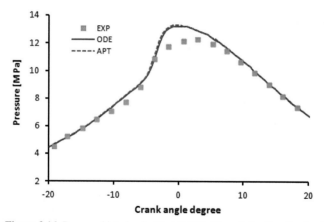

Figure 3.16: Pressure histories for Fuel E OP1 (Case 4) for Free Stochastic calculations. The mean pressure profiles of the ODE and APT models are shown with blue and red dashed lines respectively. The green symbol represents the experimental measurements.

3.1.2   Comparison of APT with ODE integration (RSC & FSC)

It will be demonstrated in this section that APT captures HCCI engine auto-ignition with the same accuracy as the direct ODE integration as far the main chemical species are concerned.

3.1.2.1   Comparisons for RSC calculations

The purpose of this section is to establish that the APT model is essentially the same as the ODE integration model. The accuracy of the cool flame (formation of formaldehyde) and main ignition (consumption of fuel) events as well as the blue flame (consumption of formaldehyde) are examined. All the four operating points are considered and their mean heat release rate for ten engine cycles are shown in Figures 3.17-3.21.

In each case the cool flame, blue flame and the main excitation markers are as accurately captured using APT as with ODE integration model. The pronounced blue flame peak for Fuel E OP1 (Figure 3.21) are well captured by the ODE integration model and APT look up. The high accuracy of APT in capturing the cool flame while using a single progress variable (including enthalpy and pressure) at each of the four operating points is an important finding. It demonstrates that the choice of progress variable (defined in terms of enthalpy of formation evaluated at 298 K) is a good one. It also suggests that there is a low dimensional manifold in the composition space for the low- temperature auto-ignition process. It should be pointed out that Intrinsic Low Dimensional Manifold (ILDM) yielded limited success for capturing the cool flame [4].

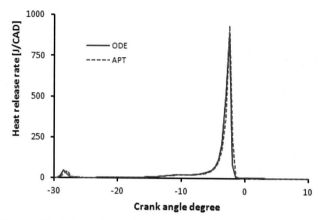

Figure 3.17: Calculated mean heat release rate [J/CAD] versus runtime in crank angle degree for Fuel D OP3 (Case 1) for RSC calculations. Engine calculations using the ODE integration and APT models are represented with blue and a red dashed line respectively.

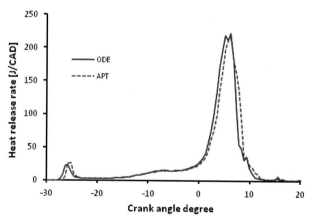

Figure 3.18: Calculated mean heat release rate [J/CAD] versus runtime in crank angle degree for Fuel C OP3 (Case 2) in the RSC calculations. Engine calculations using the ODE integration and APT models are represented with blue and a red dashed line respectively.

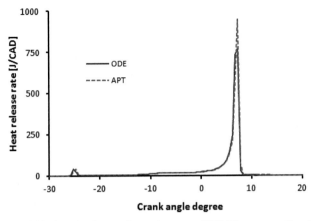

Figure 3.19: Calculated mean heat release rate [J/CAD] versus runtime in crank angle degree for Fuel C OP4 (Case 3) in the RSC calculations. Engine calculations using the ODE integration and APT models are represented with blue and a red dashed line respectively.

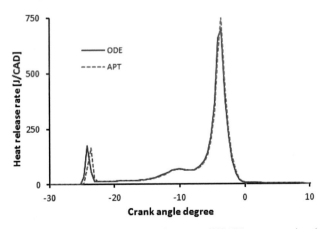

Figure 3.20: Calculated mean heat release rate [J/CAD] versus runtime in crank angle degree for Fuel E OP1 (Case 4) in the RSC calculations. Engine calculations using the ODE integration and APT models are represented with blue and a red dashed line respectively.

3.1.2.2 Comparisons for FSC calculations

The Figures 3.22-3.25 show the mean heat release rate [J/CAD] over 30 cycles for ODE integration and APT computations. In these figures, the mean heat release shows the two stage ignition characteristic of these fuels. In each case the agreement between the ODE integration and the APT models is excellent. In Figures 3.23-3.24, the main heat release peak for APT calculations deviates slightly from ODE integration predictions. This tendency is negligible in Figure 3.22 and 3.25. This is because the cases with significant cyclic variations (Figures 3.23-3.24) require higher accuracy from APT than those with negligible cyclic variations (Figures 3.22 and 3.25).

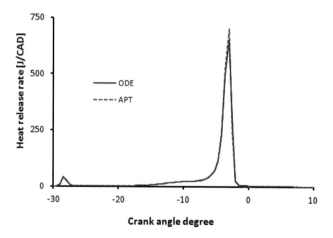

Figure 3.22: Calculated histories of the heat release rate for Fuel D OP3 for the free stochastic calculations. Engine calculations using the ODE integration and APT models are represented with blue and a red dashed line respectively.

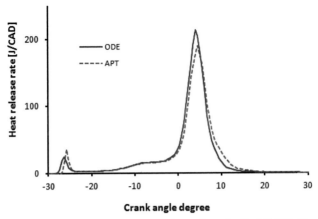

Figure 3.23: Calculated histories of the heat release for Fuel C OP3 (Case 2) for Free Stochastic calculations. Engine calculations using the ODE integration and APT models are represented with blue and a red dashed line respectively.

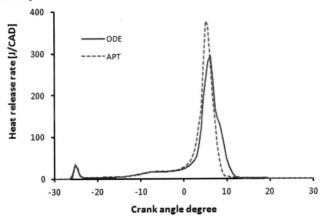

Figure 3.24: Calculated histories of the heat release rate for Fuel C OP4 (Case 3) for the free stochastic calculations. Engine calculations using the ODE integration and APT models are represented with blue and a red dashed line respectively.

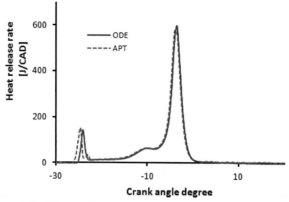

Figure 3.25: Calculated histories of the heat release for Fuel E OP4 (Case 4) for the free stochastic calculations. Engine calculations using the ODE integration and APT models are represented with blue and a red dashed line respectively.

3.1.3 Parametric study of APT

In the previous section it was shown that APT captures the main excitation event for the four engine case with the same accuracy as the ODE integration model. In this section the influence APT parameters such as ISAT error tolerance on the accuracy and local region reuse of APT is investigated. For the accuracy tests some typical cool flame (formation of formaldehyde), blue flame (consumption of formaldehyde) and main excitation (consumption of intermediates to CO_2 and H_2O) markers in *n*-heptane/toluene combustion are considered.

3.1.3.1 The influence of ISAT error tolerance on APT

The influence of ISAT error tolerance on the accuracy of APT is also investigated. The sensitive third operating point (Case 3) was considered. In Case 3 the main excitation event occurs after the Top Dead Centre. In this case combustion ends late in the expansion phase. The pressure rise from the chemical reactions competes with the pressure drop caused by the expansion of the cylinder volume. Thus, a small change in the ignition temperature can give rise to significant changes in ignition timing. This case is very sensitive and it requires a high accuracy of APT.

In Figures 3.26-3.27 the ODE integration and APT profiles for formaldehyde (CH_2O) and the lumped heptyl-ketone (L-$C_7H_{15}O$, L represents the word lumped) are shown. The species L-$C_7H_{15}O$ and CH_2O are very good cool flame markers. The spike-shaped curve of L-$C_7H_{15}O$ is accurately captured by the APT look-up table as shown in Figure 3.16. The APT curve with ISAT error tolerance equals 0.002 appears almost on the same line with the ODE integration curve despite the small magnitude of L-$C_7H_{15}O$ mass fractions. As the value of ε is increased from 0.002 to 0.01, there are slight deviations of the APT curve from the ODE integration predictions. In this case L-$C_7H_{15}O$ is produced slightly late with a higher peak mass fraction. The ridge-like structure of formaldehyde time history is reproduced by APT. Like L-$C_7H_{15}O$, the error in CH_2O when APT is used increases with the magnitude of the ISAT error tolerance. The production of CH_2O connotes the cool flame region and its consumption indicates the blue flame region. The ODE integration and APT profiles for L-$C_7H_{15}O$ and CH_2O are in very good agreement. The APT and ODE integration curves for additional important species such as CO, CO_2, HO_2 and OH are illustrated in Figures 3.28-3.31. The species CO_2 and OH are produced during the main auto-ignition event while HO_2 and CO are consumed during this phase. APT faithfully captures the ODE integration curves for CO, CO_2, HO_2 and OH. In each species the error in APT increases with the ISAT error tolerance.

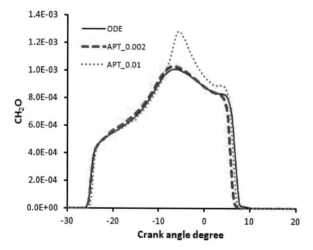

Figure 3.26: Calculated CH_2O mass fractions versus crank angle degree for Case 3. The ODE integration model is represented by a line. APT_0.002 and APT_0.01 represent APT calculations with $\square$ equals 0.002 and 0.01 respectively.

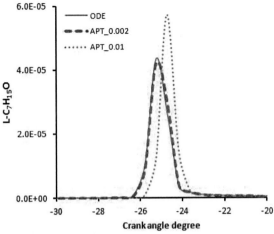

Figure 3.27: Calculated L-$C_7H_{15}O$ mass fractions versus crank angle degree for Case 3. The ODE integration model is represented by a line. APT_0.002 and APT_0.01 represent APT calculations with $\square$ equals 0.002 and 0.01 respectively.

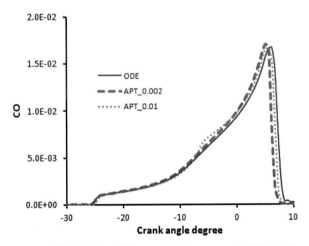

Figure 3.28: Calculated CO mass fractions versus in CAD for Case 3. The ODE integration model is represented by a line. APT_0.002 and APT_0.01 represent APT calculations with ε equals 0.002 and 0.01

respectively.

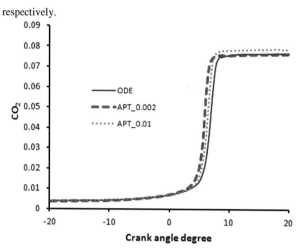

Figure 3.29: Calculated CO_2 mass fractions versus in crank angle degree for Case 3. The ODE integration model is represented by a line. APT_0.002 and APT_0.01 represent APT calculations with ε equals 0.002 and 0.01 respectively.

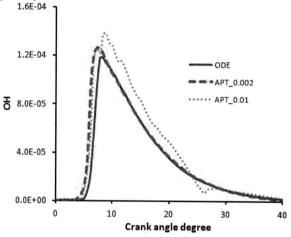

Figure 3.30: Calculated OH mass fractions versus crank angle degree for Case 3. The ODE integration model is represented by a line. APT_0.002 and APT_0.01 represent APT calculations with ε equals 0.002 and 0.01 respectively.

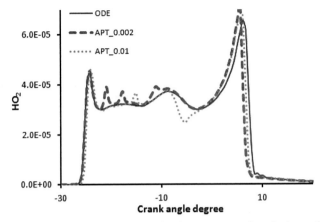

Figure 3.31: Calculated HO_2 mass fractions versus crank angle degree for Case 3. The ODE integration model is represented by a line. APT_0.002 and APT_0.01 represent APT calculations with ε equals 0.002 and 0.01 respectively.

It has been illustrated that using a single progress variable, one can accurately capture the time histories of the major and minor species as well as pollutants. APT successfully captures the cool and blue flame and as well as the main excitation markers without any loss in accuracy. In this approach no assumption was made about the detailed mechanism. Only the progress variable was specified.

3.1.3.2 The influence of APT parameters on local region reuse

The computational speed-up of APT depends on the frequency of the different local polynomial (*identical*, ISAT and PRISM) calls. With this information for a given operating point APT users will be able to decide which values of α and ε suits their application. The APT user should have in mind the level of error he/she can tolerate for the benefit of computational speed-up. The third operating point (Case 3) is selected for a single a single cycle HCCI engine calculation.

The EOA parameter β_2 was set to 120. All other APT variables remained the same. For example the number of particles required to construct second order polynomials. The parameter α is varied from 1.2 to 9.2 while ε is varied from 0.002 to 0.012. The SRM simulates 100 particles in each case. In Figure 3.32 for each α, the ratio of ISAT calls increases with ε. This is because the larger the value of ε the bigger ISAT EOA, more initial conditions will encounter ISAT EOAs. This gives rise to more ISAT calls. Although the highest number of ISAT calls were recorded for very small values of α. This is because for very small α means the disallowed region of the PRISM hypercube is larger than the allowed region, as such more ISAT EOA reuse in the disallowed region. This increases the overall frequency of ISAT calls as shown in Figure 3.32

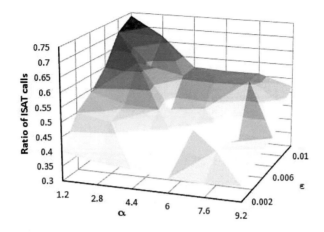

Figure 3.32: The variation of ISAT calls with ISAT error tolerance and the parameter α for engine Case 3. The magnitude in the ISAT calls axis increases with the intensity of the blue color.

The doom structure in Figure 3.33 shows the ratio of PRISM calls for different values α and ε. No PRISM calls are observed for small values of α were used (α close to zero), because the allowed section of the PRISM hypercube is infinitesimal. As α increases from small values, for each value of ε, the number of PRISM calls increases very fast with α. This reaches a maximum at about α = 7.0. Beyond this value α = 7.0, the ratio of PRISM calls almost flattens out. It could be expected that the highest number of PRISM calls is found for the biggest values of α and ε.

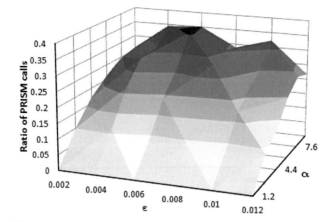

Figure 3.33: The variation of PRISM calls with ISAT error tolerance and the parameter α for engine Case 3. The magnitude in the PRISM calls axis increases with the intensity of the blue color.

The plot ratio of *identical* calls versus α and ε is shown in Figure 3.34. The ratio of identical calls does not depend strongly on α. For large ε, the size of the ISAT EOA increases as well as the size of the identical EOA. Therefore, there are more *identical* calls for large values of ε. As shown in Figure 3.35, the frequency of growth calls decreases as α increases. There are also more growth calls for smaller values of the ISAT error tolerance as shown in Figure 3.35. Large values of α imply larger allowed section of the PRISM hypercube. Second order polynomial approximation reuses are more likely for hypercubes with second order polynomial coefficients. Small values of ε give small ISAT EOAs, thus more initial condition will enter the neighborhood of the ISAT EOAs and the frequency of growth calls will increase.

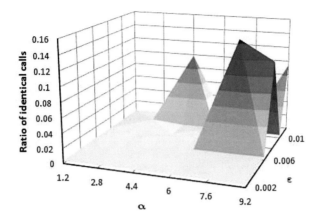

Figure 3.34: The variation of *identical* calls with ISAT error tolerance and the parameter α for engine Case 3. The magnitude in the *identical* calls axis increases with the intensity of the blue color.

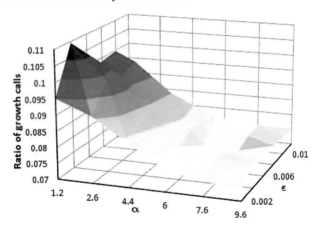

Figure 3.35: The variation of growth calls with ISAT error tolerance and the parameter α for engine Case 3. The magnitude in the growth calls axis increases with the intensity of the blue color.

3.1.4 Control of tabulation errors in APT

Additional tests were performed with the SRM-HCCI engine to analyze the APT tabulation errors. Sources of errors are the approximations through first and second order polynomials, and the modeling of the combustion process with a single progress variable. The third operating point (Case 3) was used for the tests because of its sensitivity to tabulation errors. All calculations were performed with 100 particles. In each APT polynomial approximation, the direct ODE integration code is invoked also to calculate the local error in APT for each time step.

$$\xi = \frac{1}{N''} \sum_{i=1}^{N''} \left(\left| \frac{\phi_{i,APT}^{t+\Delta t} - \phi_{i,ODE}^{t+\Delta t}}{\phi_{i,ODE}^{t+\Delta t}} \right| \right) \tag{3.1}$$

Where $\phi_{APT}^{t+\Delta t}$ and $\phi_{ODE}^{t+\Delta t}$ are the ODE solutions computed with the APT and the ODE integration model respectively and N'' is the number of chemistry queries for the given time step. This way the local error in the approximation can be calculated, and compared to the given ISAT error tolerance, which was set to 0.002 and α was set to 7.6. It should be noted that α is proportional to the size of the allowed section of the PRISM hypercube. Errors that are higher than the given ISAT tolerance are caused by the usage of a single progress variable. As shown in Figure 3.36, at least 99.4 % of ISAT calls have errors less than the ISAT error tolerance. In Figure 3.37 it is shown that at least 98.2 % of PRISM calls have errors less than the ISAT error tolerance. The largest PRISM error is 9 times the ISAT error tolerance while the largest ISAT error is 10 times the ISAT error tolerance as illustrated in Figures 3.36 and 3.37.

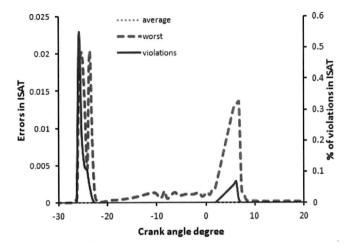

Figure 3.36: Test of local ISAT error control plotted against time in crank angle degree: the average error calculated; the worst error made to date; the percentage of ISAT calls that causes a tolerance violation, error greater than the ISAT error tolerance. The APT parameters used were: $\varepsilon = 0.002$ and $\alpha = 7.6$.

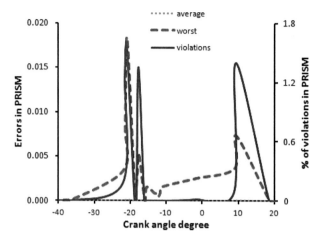

Figure 3.37: Test of local PRISM error control plotted against time in crank angle degree: the average error calculated; the worst error made to date; the percentage of PRISM calls that causes a tolerance violation, error greater than the ISAT error tolerance. The APT parameters used were: $\varepsilon = 0.002$ and $\alpha = 7.6$.

3.1.5 Computational speed up factor

3.1.5.1 Speed up factor versus APT parameters

The computational performance of APT is investigated for Case 2. The computational speed-up is investigated for different values of ISAT error tolerance (ε) while α is kept constant ($\alpha = 4$) for SRM calculations with 100 particles. These calculations were performed for only one engine cycle. As shown in Figure 3.38, the computational speed up is directly proportional to ISAT error tolerance. The speed-up factor curve is close to straight line with positive gradient. When the ISAT error tolerance is larges large, it means that size of the ISAT EOA is large. The size of the ISAT

EOA is directly proportional to the size of the PRISM and identical EOA. Therefore, APT will encounter more local region reuse. This gives rise to the computational speed-up. In these calculations for every initial condition stored, 3 extra ODE integration calculations are required for the calculation of the mapping gradient. The cost of constructing second order polynomial coefficients from real initial conditions is relatively small compared to the cost of calculating the mapping gradients (see Table 3). When α is increased while ISAT error tolerance is held constant ($\varepsilon = 0.002$), the size of the allowed section each PRISM hypercube is increased, thereby increasing the frequency of PRISM calls and hence the computational speed up as shown in Figure 3.39.

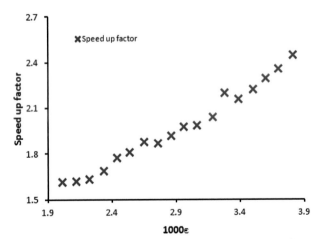

Figure 3.38: The variation of computational speedup with error tolerance ε ($\alpha = 4$) for the first cycle for Case 2.

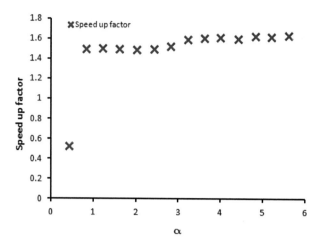

Figure 3.39: The variation of computational speedup with α, ($\varepsilon = 0.002$) for the first cycle for Case 2.

In Figures 3.38 and 3.39, when the error tolerance ε and α were varied, the computational speed-up did not exceed 2 for calculations with 100 particles. This is because based on this number of particles the number of chemistry calls cannot yield enough APT library reuse.

Table 3: Cost of various operations (normalized to the cost of one ODE solution) in APT

Operation	Normalized Cost
1 ODE solver call	1.0
1 *identical* evaluation (plus searching time)	0.00025
1 ISAT evaluation (plus searching time)	0.00037
1 PRISM evaluation (plus searching time)	0.0013
1 PRISM construction	0.021
1 mapping gradient (ISAT) construction	3.0

70

3.1.5.2 Speed up factor for RSC calculations

In order to assess the computational performance of APT, the chemistry computational time is calculated for 10 engine cycles when 100 particles are simulated for the four test cases. The time spent in the chemistry subroutines is recorded using the intrinsic FORTRAN function SYSTEM_CLOCK. In each case, the control calculations using the ODE integration are also performed and the calculation time was recorded.

In Figures 3.40–3.43, the chemistry computational speed up factor is shown for the four different cases. In each of the plots, there is almost no computational speed up factor for the first cycle. The computational speed up of 2 is obtained for the second cycle and it increased to about 10 for the third. After the fifth cycle, the APT computational speed up reaches 3 orders of magnitude and it continues in this scale right to the tenth engine cycle. For example, for Fuel D OP3, the APT and ODE integration computational costs for the first cycle are 1840.2 s and 1548.5 s respectively. The APT and ODE computational cost for the fifth cycle are 1.15 s and 1548.5 s. The computational speed up factor for the fifth cycle is 1346.5. This gives a computational speed up factor is 3 orders of magnitude. After the fifth cycle the APT computational time speed up factor is maintained to the last engine cycle with very small fluctuations introduced by the loading of the computer. The ODE integration computational time for all the cycles is about the same with small fluctuations introduced by the loading of the computer.

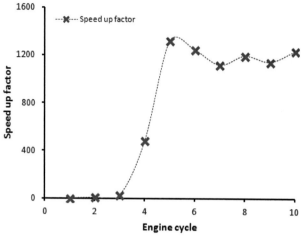

Figure 3.40: Speed up factor for Fuel D OP3 (Case 1) for RSC calculations.

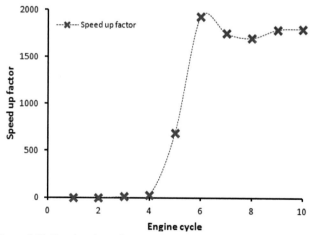

Figure 3.41: Speed up factor for Fuel C OP3 (Case 2) for RSC calculations.

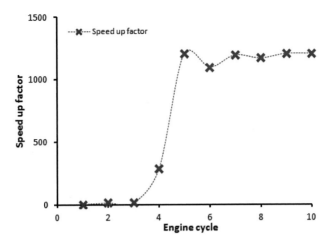

Figure 3.42: Speed up factor for Fuel C OP4 (Case 3) for RSC calculations.

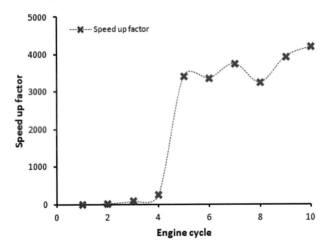

Figure 3.43: Speed up factor for Fuel E OP1 (Case 4) for RSC calculations.

3.1.5.3 Speed up factor for FSC calculations

In Figures 3.44-3.47 the computational time for 30 individual engine cycles are shown. For the first cycle, the APT computational expenditure is about the same as that of the direct ODE integration, therefore no computational speed up is recorded. This is because during the first cycle the initial APT table is being constructed. This table comprises the mapping gradients, initial condition information and second order polynomial coefficients for each PRISM hypercube. The ratio of the average cost of mapping gradient calculation and one ODE integration call is 3. In this simulation the polynomial calls could barely recoup the cost of mapping gradient and second degree polynomial construction. In the subsequent cycles, the probability of mixture initial conditions to encounter PRISM hypercubes, ISAT and *identical* EOA with similar composition increases. Therefore more polynomial reuses are counted and the computational speed up factor increases accordingly. In Figures 3.45-3.46 the APT computational speed up factor increased as engine cycles increased in a fluctuating pattern. This tendency is because of the strong cyclic variations in these test cases which require high accuracy of APT. In Figures 3.44 and 3.47 the APT computational speed up factor increased as engine cycles increased with negligible fluctuations. This is because these test cases possess less cyclic variations. Computational speed ups exceeding 12 were obtained. In this work, memory requirement for each operating points was not investigated.

74

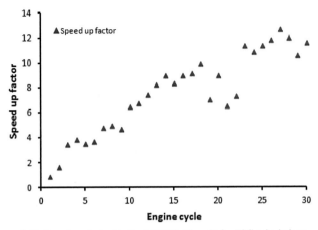

Figure 3.44: Speed up factor for Fuel D OP3 (Case 1) for FSC calculations.

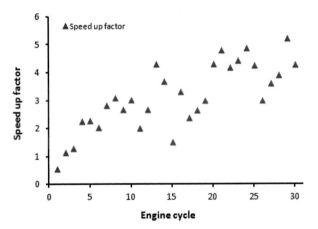

Figure 3.45: Speed up factor for Fuel C OP3 (Case 2) for FSC calculations.

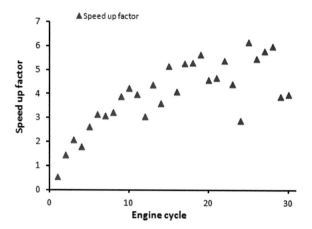

Figure 3.46: Speed up factor for Fuel C OP4 (Case 3) for FSC calculations.

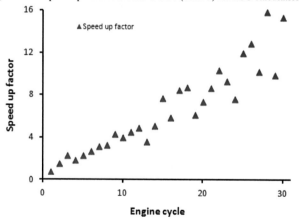

Figure 3.47: Speed up factor for Fuel E OP1 (Case 4) for FSC calculations.

3.1.6 Reduced library APT model calculations

3.1.6.1 Nth cycle comparisons

The reduced library APT model is used in this section to study the most important reactions and species present in n-heptane/toluene mechanism. A skeletal n-heptane/toluene mechanism that was developed using chemistry guided reduction technique was used [19]. The accuracy of species that participate in the most sensitive reactions in this mechanism is studied. These include O, H, OH, $C_6H_5CH_2$ and C_2H_3 and additional species. The species O, H and OH participates in the main chain branching reactions. The species $C_6H_5CH_2$ is important for resistance to auto-ignition and aromatic species formation respectively. The species C_6H_5 involved in benzene oxidation was also studied [19].

The potential of APT in capturing these sensitive reactions and species in multi-cycle HCCI engine simulations is studied. In this case the modified algorithm of APT was used with a limited number of ODE integration calls per time step. At most 15 stored initial were stored per time step for the first five cycles. After the fifth cycles, the cumulative sum of initial conditions stored for all the cycles per time should not exceed 15. For simplicity the 25th cycle was selected, because it recorded the highest computational speed up factor (approximately 16). As shown in Figures 3.48-3.51, the species (H, OH, O and O_2) that participate in the most sensitive reactions (having the strongest influence on ignition timing [19]) were accurately captured by APT.

The APT profiles for $C_6H_5CH_2$ and $C_6H_5CH_3$ and C_6H_5 are in excellent agreement with those of ODE integration as shown in Figures 3.52-3.54. The unsaturated radical C_2H_3 and the species (HCCO, C_2H_4 and C_2H_2) present in the baseline sub-mechanism are plotted in Figures 3.55-3.59. The APT profiles for each of the species were in very good agreement with the ODE integration profiles. It shows that the choice of progress variable is a good one and using a limited number of stored initial conditions per time step improves the efficiency of the algorithm.

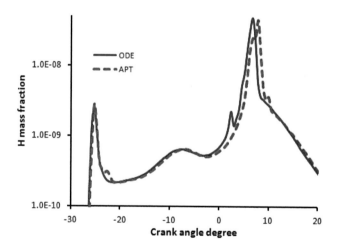

Figure 3.48: Calculated H mass fractions for the 25th cycle versus runtime in crank angle degree for Case 3.

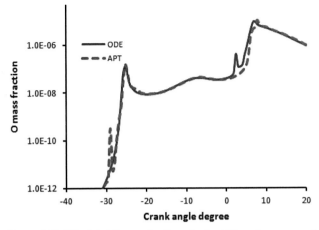

Figure 3.49: Calculated O mass fractions for the 25th cycle versus runtime in crank angle degree for Case 3.

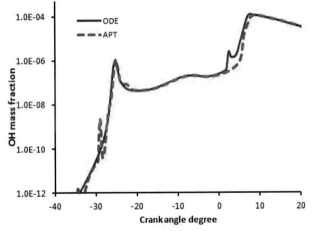

Figure 3.50: Calculated OH mass fractions for the 25th cycle versus runtime in crank angle degree for Case 3.

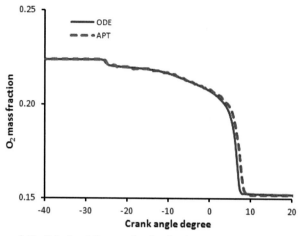

Figure 3.51: Calculated O_2 mass fraction for the 25th cycle versus runtime in crank angle degree for Case 3.

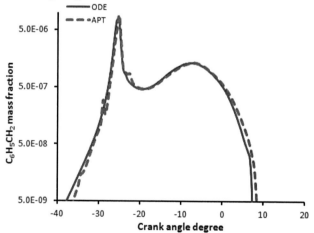

Figure 3.52: Calculated $C_6H_5CH_2$ mass fractions for the 25th cycle versus runtime in crank angle degree for Case 3.

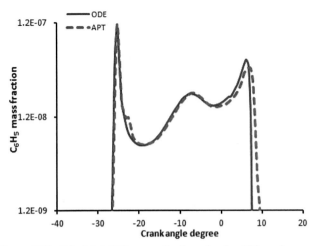

Figure 3.53: Calculated C_6H_5 mass fractions for the 25th cycle versus runtime in crank angle degree for Case 3. The ODE integration model is represented by a line and the APT model is denoted by a dashed line.

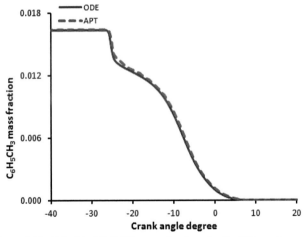

Figure 3.54: Calculated $C_6H_5CH_3$ mass fractions for the 25th cycle versus runtime in crank angle degree for Case 3. The ODE integration model is represented by a line and the APT model is denoted by a dashed line.

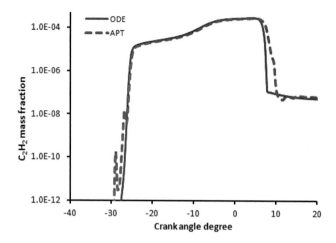

Figure 3.55: Calculated C_2H_2 for the 25th cycle versus runtime in crank angle degree for Case 3.

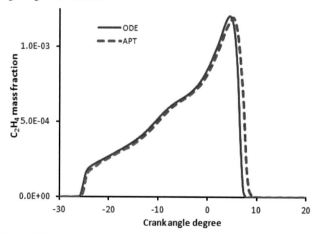

Figure 3.56: Calculated C_2H_4 mass fractions for the 25th cycle versus runtime in crank angle degree for Case 3.

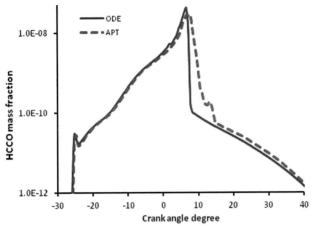

Figure 3.57: Calculated HCCO mass fractions for the 25th cycle versus runtime in crank angle degree for Case 3.

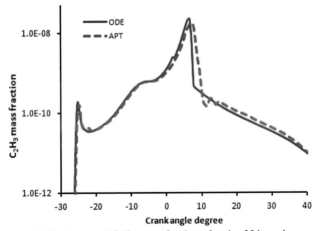

Figure 3.58: Calculated C_2H_3 mass fractions for the 25th cycle versus runtime in crank angle degree for Case 3.

3.2.6.2 Comparisons for the mean profiles

In this section, the ODE and APT mean profiles for temperature, heat release, and some important species (CH_2O, CO and CO_2) are compared. These mean values have been computed from 30 SRM HCCI engine cycles. As shown in Figure 3.59-3.60 the temperature and heat release profiles demonstrate two peaks representing the cool flame and the main excitation step. The heat release profile has a small elbow between the cool flame and the main excitation. This is the blue flame region. It connotes the consumption of CH_2O. There are very small deviations between ODE integration and APT models for the mean temperature profiles. The mean ODE integration heat release falls on the same line as the APT model for the cool flame and blue flame regions, but there are slight deviations around the main excitation regions.

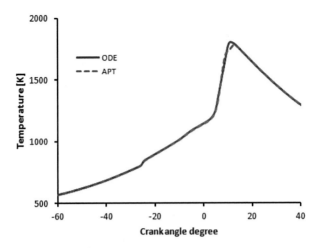

Figure 3.59: Calculated histories of the mean temperature for Case 3. The means were calculated from 30 cycles.

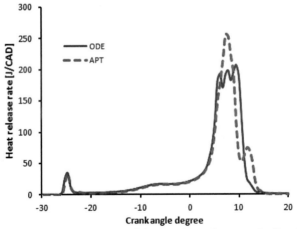

Figure 3.60: Calculated histories of the mean heat release rate for Case 3. The means were calculated from 30 cycles.

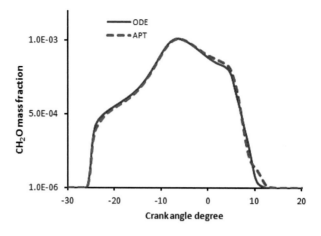

Figure 3.61: Calculated histories of the mean CH_2O for Case 3. The means were calculated from 30 cycles.

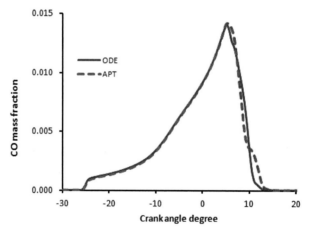

Figure 3.62: Calculated histories of the mean CO for Case 3. The means were calculated from 30 cycles.

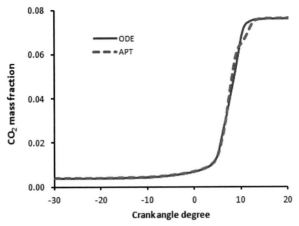

Figure 3.63: Calculated histories of the mean CO_2 for Case 3. The means were calculated from 30 cycles.

For the intermediate species (CO and CH_2O) these deviations around the main excitation region are more visible. For the mean species (CO, CO_2 and CH_2O) histories shown in Figures 3.61-3.63, the agreement between SRM calculations with ODE integration and APT model is very good.

3.2.6.3   Tabulation errors for the reduced library APT model

In this section the error associated with APT is presented. For each ISAT and PRISM call, the ODE integration solver is called and the local error is calculated using the Equation (3.1). The errors in ISAT and PRISM approximations for the 2nd, 6th and 25th engine cycles are shown in Figures 3.64 and 3.65 respectively. The errors are calculated as the average over all queries to date for the given time step. The violation is calculated as the percentage of queries to date for which the local error exceeds the ISAT error tolerance for the given time step.

The magnitude of the ISAT error increases fairly with runtime in crank angle degrees. This is due to the propagation of round up errors. The ISAT errors do not always increase as the calculation progresses from one cycle to the next. The ISAT error will depend on the degree of deviation of the compositions of the new cycle as compared to those stored in the library. As shown in Figure 3.64 fewer operator splitting time steps were involved in the calculation of the ISAT errors for Cycles 6 and 25, because PRISM reuse dominated in most of the cycles beyond Cycle 5. It is shown in Figure 3.65 that Cycle 25 with the highest speed up factor has the smallest PRISM error. In these three cycles shown, their PRISM errors increase with runtime because of accumulation of round up errors.

In Figure 3.67, the percentage of ISAT violations decreases with runtime in crank angle degree, while that of PRISM (in Figure 3.68) increases with runtime in crank angle degree. The sixth cycle displayed the least frequency of PRISM calls but it had the highest percentage of violations. The violations in PRISM depend strongly on the APT library details. If the APT library was constructed with compositions that differ greatly from those of the trajectories of Cycle 6, the magnitude of the PRISM violations for cycle 6 will be larger.

88

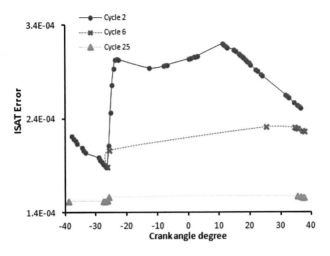

Figure 3.65: Test of local ISAT error control. Plotted against time in crank angle degree: the average error calculated; blue diamond represents cycle 2; red circle represents cycle 6 and crosses represents cycle 25 ($\varepsilon = 0.0025$ and $\alpha = 6$).

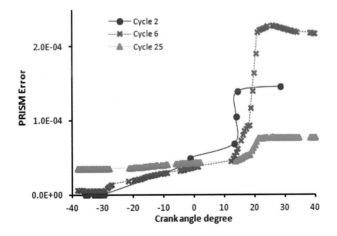

Figure 3.66: Test of local PRISM error control. Plotted against time in crank angle degree: the average error calculated; blue diamond represents

89

cycle 2; red circle represents cycle 6 and crosses represents cycle 25 (ε $\square$= 0.0025 and α = 6).

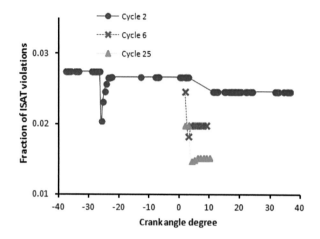

Figure 3.67: Plotted against time in crank angle degree: the percentage of ISAT calls that causes a tolerance violation; blue diamond represents cycle 2; red circle represents cycle 6 and crosses represents cycle 25 (ε = 0.0025 and α = 6).

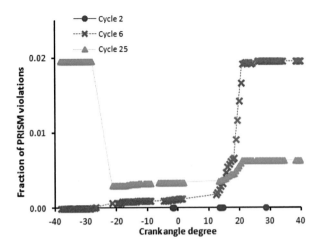

Figure 3.68: Plotted against time in crank angle degree: the percentage of ISAT calls that causes a tolerance violation; diamond represents cycle 2; red circle represents cycle 6 and crosses represents cycle 25 ($\varepsilon = 0.0025$ and $\alpha = 6$).

3.1.6.4 Computational speed up with reduced library APT model

In this section the computational speed up associated with the reduced library APT model is presented. As shown in Figure 3.69, a computational speed up factor of 16 was obtained for SRM calculations with cyclic variations for Case 3. In contrast to the computational speed up factor illustrated in Figure 3.46, this modification of APT resulted in improvement in the computational speed up greater than a factor of 2.

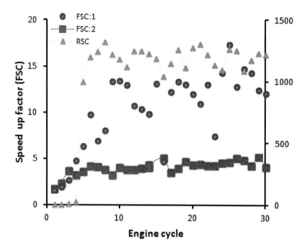

Fig. 3.69: The variation of computational speed up with engine cycle Free Stochastic calculations (FSC). FSC 1 represents the calculations with this new version of APT. FSC-2 represents the calculations with original APT code. RSC represents the Repeated Single Cycle calculations.

The storage of a limited number of initial conditions per time improves the computational gain factor of APT as extra ODE integration calculations for mapping gradient construction are avoided.

Chapter 4   Conclusion and future outlook

In this thesis APT has been presented and thoroughly described. It parameterizes the solution of the chemical rate equation system with zeroeth, first order and second order algebraic expressions. The algebraic polynomials are computationally inexpensive to evaluate as compared to direct ODE integration. To facilitate the searching of initial conditions, the chemical composition space is partitioned into equally size hypercubes. The searching and parameterization is performed in terms of only three variables. These include: standard enthalpy formation evaluated at 298 K, pressure and total enthalpy. In real time each hypercube is divided into adaptive hypercubes. The hypercube and adaptive hypercubes are accessed via linear and binary trees. The initial conditions used for the construction of second order polynomials are stored in the adaptive hypercubes. Two ellipsoids are drawn around each stored initial condition. The inner and outer ellipsoids are the regions for zeroeth order and first polynomial approximations to the ODE solutions respectively.

When used with n-heptane/toluene fueled SRM-HCCI engine simuations, APT yielded significant computational speed-up without any significant loss in accuracy. For multi-cycle calculations with the same random seed, computational speed up exceeding 3000 was obtained for the fourth operating point. Therefore, in this situation APT demonstrated higher computational speed up factor than that obtained with ISAT (speed up factor of 1000 [7]) and PRISM (speed up factor of 10 [8]). The cool

93

flame, blue flame and main excitation events were accurately captured for all the four operating points. A computational speed up factor of 12 was obtained for multi-cycle SRM calculations with different random seed. The cool flame, blue flame and main excitation events were accurately captured for all the four operating points. The second and third operating points demonstrated high sensitivity; their ignition events occurred after Top Dead Centre and these events were also captured by APT. These cases showed larger variations in ignition delay times from one cycle to the next and APT successfully captured the cyclic variations without significant loss in accuracy.

It is demonstrated that number of second degree polynomial reuse depends strongly on the size of the allowed section of the adaptive hypercubes. The first order polynomial reuse depends strongly on the ISAT error tolerance. About 99 % of the first order polynomial reuse has errors less than the ISAT error tolerance. The maximum error in the first order polynomial approximation is greater than that for second order polynomials.

As compared to PRISM, rather than using the entire set of chemical species to perform searching and parameterization, APT replaces the entire set of (148 chemical species as used in this thesis) with a progress variable. Therefore, a drastic reduction in the memory requirement and improved computational speed up and faster searching times. Initial conditions are stored only outside the EOA of stored initial conditions. This enhances spread of stored initial condition as they enter the PRISM hypercube. The spread of initial conditions is also improved by the growth of the ISAT EOA. The degree of spreading of the stored initial conditions is directly proportional to the accuracy of the second order polynomial coefficients. However, with this algorithm of using real stored initial conditions it is not possible to obtain the degree of spreading of initial conditions provided central composite design that was used in the original PRISM formulation.

In APT second order polynomials are not created for every hypercube when a mixture initial condition enters it for the first time as in PRISM. As the reacting trajectory progresses through a PRISM hypercube in APT, zeroeth and first order polynomial are used to approximation the ODE solutions for initial conditions within an ISAT EOA. This process continues and more local region reuses are registered improving the computational efficiency of code until enough initial conditions are stored for second order polynomial construction. Therefore in APT it is not possible to have a PRISM hypercube with limited or no reuse as in PRISM.

As compared to the ISAT method, APT is not limited to first order polynomial approximations but it performs zeroeth, first and second order polynomials as the need arises. The polynomial construction takes place only when the need arises. Searching time is relatively smaller in APT as compared to ISAT. APT algorithm circumvents for long binary trees by building separate binary trees on the data structure of each hypercube. This makes the search for the closest initial condition to a given initial condition a simple and local task.

Contrary to APT, PRISM constructs the second order polynomial expressions for a given hypercube by calculating the ODE solution via direct ODE integration at specified locations in the hypercube determined by central composite design. The extra computational cost compromises the computational efficiency of PRISM. Each second order polynomial must be used at least 250 times before its computational cost is recouped. In APT the cost of constructing the second order polynomials is only about 2 % of the cost of one ODE integration call. This is because APT uses real initial conditions that have been stored from previous calculations. PRISM lacks the control of tabulation errors. The sizes of the adaptive hypercubes are calculated from the ISAT Ellipsoid of Accuracy in APT. The ISAT EOAs are calculated from the mapping gradient.

In the SRM model for HCCI engines, the progress variable and total enthalpy are considered as random variables whereas pressure is not. All particles per time step have the same pressure, therefore there is limited spread in the pressure axis. As an attempt to circumvent for this limitation, in this thesis a reduced library APT model was developed and tested. At each time step at most 15 initial conditions are stored for the first five engine cycles. After the fifth engine cycle, initial conditions are added if the cumulative number of initial conditions stored after the fifth cycle is less than 15 for the given time step. The benefits of this approach include: firstly, it gives requires smaller memory and an improved spread of initial condition especially in the pressure direction, it gives more accurate second

96

order polynomials and eliminates the redundant ODE integration calculations for the construction of the mapping gradients

It was shown that with these adjustments, APT is capable of capturing the main species that participate in the main reactions (baseline chemistry, benzene and toluene) of n-heptane/toluene combustion. In the multi-cycle calculations, APT is capable of capturing the cool flame and main excitation markers without any significant loss in accuracy. The computation gain factor took values between 4 and 16 for the SRM calculations using this reduced library APT model.

However, despite the improvements of APT presented in this work as compared to PRISM and APT, the initial conditions within a PRISM hypercube are constrained by the unity of the sum of mass fractions of each its initial condition. Not every position in the PRISM hypercube has a feasible value for the species mass fractions. Therefore an approximation can easily go off the response surface constructed and reaction trajectory.

The possibility of increasing the accuracy of the second order polynomials coefficient by adding a couple of points close each face of the PRISM hypercube cannot be accomplished for the same reason. Also to include initial conditions from the eight neighboring hypercube outside each of its faces might be challenging, because some PRISM hypercubes near edges may not have 8 neighbours. For those with 8 nearest neighbors, the neighboring hypercubes may not have stored in stored initial conditions.

For each PRISM hypercube the initial condition slope data store include: N_p initial conditions (3 variables each), N_p ODE solutions (n_s+1 variables each), N_p sensitivities of the n_s chemical species with respect to progress variable, total enthalpy and pressure ($3n_s$ each), and N_p sensitivities of progress variable, total enthalpy and pressure with respect to their initial conditions (9 variable each). Memory for each PRISM hypercube in APT could be optimized if initial condition-slope data is deleted immediately after the construction of second order polynomial coefficients. This deletion process has the ability to reduce the PRISM hypercube memory by a factor greater than 50. However, in this test with current APT code, deletion of this initial condition slope data after construction of second order polynomials cannot be accomplished. If deletion was implemented with this n-heptane/toluene mechanism, the memory for the APT storage table should have reduced by a factor greater than 80.

The accuracy of the second order polynomial coefficients can be improved by including mapping gradients information when the coefficients are constructed. Another benefit to this approach is fewer second order polynomial coefficients, thus, reduced memory requirement. This will entail the introduction of additional assumptions, that is, the mean of the stored initial conditions will be replaced by its closest neighbour. This will be demonstrated in our future work.

The test for multi-cycle SRM-HCCI engine calculations were limited to 100 computational particles on the premise of reduced computational requirement which favors model development, although real PDF applications may involves several thousands of computational particles. However, this is the first time that real time second order polynomial expressions have been used for representing chemistry in engine simulations and it has been implemented for a complex chemical mechanism with more 100 chemical species.

A mechanism of n-heptane/toluene with 148 species and 1281 elementary reactions was used in this work. The target is larger hydrocarbon mechanisms used in diesel combustion such as n-decane. In this case, APT will be combined with a detailed reduction method (QSSA [1]) and an elegant dimension reduction tool, Invariant Constrained Equilibrium Pre-image Curve (ICE-PIC) method [6]. Initially the mechanism will be reduced by QSSA, then, the reduced mechanism will be send to the APT/ICE-PIC coupling. This will also be demonstrated in our future work.

Chapter 6 References

1   Peters, N. and Williams, F. A., The asymptotic structure of stoichiometric methane air flames, Combust. Flame, 68, 185-207, 1987.

2   Keck, J. C. and Gillespie, D., Rate-controlled partial-equilibrium method for treating reacting gas mixtures, Combust. Flame, 17, 237-241, 1971.

3   Ahmed, S. S., Mauss, F., Moreac, G. and Zeuch, T., A comprehensive and compact n-heptane oxidation model derived using chemical lumping, Phys. Chem. Chem. Phys., 9, 1107-1126, 2006.

4   Maas, U. and Pope, S., Simplifying chemical kinetics: intrinsic low dimensional manifolds, Combust. Flame, 88, 239-264, 1992.

5   Lam, S. H., Using CSP to understand complex chemical kinetics, Combust. Sci. Technol., 89, 375–404, 1993.

6   Ren, Z., Pope, S. B., Vladimirsky, A. and Guckenheimer, J. M., The invariant constraint equilibrium pre-image curve method for dimension reduction of chemical kinetics, The J. Chem. Phys., 124, 14111, 2006.

7   Pope, S. B., Computationally efficient implementation of detailed chemistry using in situ adaptive tabulation, Combust. Theory & Modelling, 1, 41-63, 1997.

8   Tonse, S., Moriarty, N. Brown, N. and Frenklach, M., PRISM: Piecewise Reusable Implementation of Solution Mapping, Isr. J. Chem., 39, 97-106, 1999.

9   Christo, F. C., Masri, A. R., Nebot, E. M. and Pope, S., An integrated PDF/neural network approach for simulating turbulent reacting systems, Proc. Combust. Inst., 43-48, 1996.

10 Lehtiniemi, H., Mauss, F., Balthasar, M. and Magnusson, I., Modelling Diesel spray Ignition using detailed chemistry with a flamelet progress variable approach, Combust. Sci. Tech., 178, 1977-1997, 2006.

11 Mauss, F., Ebenezer, N. and Lehtiniemi, H., Adaptive Polynomial Tabulation (APT): A computationally economical strategy for the HCCI engine simulation of complex fuels, SAE paper, 2010-01-1085, 2010.

12 Box G. and Draper, N., Empirical model-building and response surfaces, John Wiley and Sons, New York, 1987.

13 Maigaard, P., Mauss, F. and Kraft, M., Homogeneous Charge Compression Ignition engine: A simulation study of the effects of inhomogeneities, ASME, J. Eng. Gas

Turbines Power, 125, 466-471, 2003.

14 Bhave, A., Kraft, M., Montorsi, L. and Mauss, F., Modelling a dual fuelled multi-cycle engine using PDF-based engine cycle simulator, SAE-2004-01-0561, 2004.

15 Bhave, A., Balthasar, M., Kraft, M. and Mauss, F., Analysis of natural gas fuelled Homogeneous Charge Compression Ignition engine with Exhaust Gas Recirculation using Stochastic Reactor Model, The Int. J. Eng. Research, 5(1), 93-104, 2004.

16 Tuner, Martin, Stochastic Reactor Model for engine simulations, Doctoral Thesis at Lund University, ISBN 978-91-628-7416-2, 2008.

17 Herzler, J., Fikri, M., Hitzbleck, K., Starke, R., Schulz, C., Roth, P. and Kalghatgi, G., Shock-tube study of the autoignition of n-heptane/toluene/air mixtures at intermediate temperatures and high pressures
Combust. Flame, 1-2,25-31, 2007.

18 Kalghatgi, G., Risberg, P. and Angstrom, H-E., A method of defining ignition quality of fuels in HCCI engines, SAE2003-01-1816, 2003.

19 Ahmed, S., Mauss, F. and Zeuch, T., The generation of a compact n-heptane/toluene using chemistry guided reduction (CGR) technique, Z. Phys. Chem., 223, 551-563, 2009.

20 Zeuch, T., Moreac, G., Ahmed, S. and Mauss, F., A comprehensive skeletal mechanism for the oxidation of n-heptane generated by chemistry-guided reduction
Combust. Flame 155, 651- 674, 2008.

21 Jollife, I. T., Principal component analysis, Springer series in statistics, second edition, New York, 2002.

22 Bellanca, R., BlueBellMouse: A tool for kinetic model development, Doctoral Thesis, Lund University, Lund, ISSN 1102-8718, 2004.

Nomenclature

Symbol	Meaning
a_{ij}	Second order polynomial coefficients
$a_{i,jk}$	Second order polynomial coefficients
$A(\phi_0)$	Mapping gradient at ϕ_0 at a time t_0
A_0	Initial condition of mapping gradient
A''	Cross sectional area
$\tilde{B}$	Second order matrix created from stored initial conditions
$\hat{B}$	Matrix of stored initial conditions
$\check{B}$	PCA reduced matrix of stored initial conditions
$\tilde{B}^T$	Transpose of $\tilde{B}$
$\overline{B}$	Product of $\tilde{B}^T$ and $\tilde{B}$
c	Progress variable
c_v	Specific heat capacity at constant volume[Joules per Kg per Kelvin]
C_ϕ	Proportionality constant
F_ϕ	Mass density function
h	Total enthalpy
h'	Fluctuation in temperature
h_g	Convective heat transfer coefficient
H_{298}	Enthalpy of formation evaluated at 298 K
J	Jacobian matrix
I	Identity matrix
k	Number of variables in a factorial design
K	Kelvin unit of temperature
p'	Fractional part of the factorial design
p	Pressure in pascal
n	Number of stored records in an ISAT look up table
N_p	Number of stored initial conditions required to create second order polynomial coefficients
n_s	Number of chemical species
Q_i	Change in mass density function due chemical reaction and change of volume
S	Chemical source term

t	Time in seconds
T_w	Wall temperature
t_∞	Time at the end of combustion
t_0	Time in seconds
U	Convective heat loss term
V	Vector of cutting plane
Y_i	Species mass fractions
$\widehat{Y}$	Matrix of stored ODE solutions
$\overline{Y}$	Product of $\tilde{B}^T$ and $\widehat{Y}$
v^T	Transpose of v

Greek letters	**meaning**
ε	ISAT error tolerance
ξ	Error
λ	Mixture strength
β_c	Curl model constant
β_1	Real constant,
β_2	Real constant,
$\delta\phi_0^0$	*identical* EOA size
$\delta\phi_0^2$	PRISM EOA size
ϕ	Initial condition at time t
φ	Sample space representation of a random variable
ϕ_q^r	Reduced query initial condition
$\phi_{APT}^{t+\Delta t}$	ODE solution from APT at ϕ^t
$\phi_{ODE}^{t+\Delta t}$	ODE solution from direct integration at ϕ^t
ϕ_q	A query initial condition
ϕ_s^r	Stored initial condition
γ	Scalar of cutting plane
ϕ_0	Initial condition at time t_0
ϕ_i	Arbitrary initial condition
ϕ_0^{out}	ODE Solution of ϕ_0 after time integration
$\delta\phi_0^1$	ISAT EOA size
Δt	Time step size in seconds

Abbreviations

ALDM	Attractive Low Dimensional Manifold
APT	Adaptive Polynomial Tabulation
BDF	Backward Difference Formula
CAD	Crank Angle Degree
CFD	Computational Fluid Dynamics
CSP	Computational Singular Perturbation
EGR	Exhaust Gas Recirculation
EOA	Ellipsoid of Accuracy
EVO	Exhaust Valve opening
FPI	Flame Prolongation of ILDM
HCCI	Homogeneous Charge Compression Ignition
ICE-PIC	Invariant Constraint Equilibrium Edge Pre-image Curve
ILDM	Intrinsic Low Dimensional Manifold
ISAT	In situ Adaptive Tabulation
IVC	Inlet Valve Closure
LU	Lower Upper
MDF	Mass Density Function
ODE	Ordinary Differential Equation
PaSPFR	Partially Stirred Plug Flow Reactor
PCA	Principal Component Analysis
PDF	Probability Density Function
PFR	Plug Flow Reactor
PSR	Perfectly Stirred Reactor
RCCE	Rate-Controlled Constraint Equilibrium
QSSA	Quasi Steady State Approximation
SI	Spark Ignition
SRM	Stochastic Reactor Model
TDC	Top Dead Center

List of Tables